Fire from First Principles

A design guide to building fire safety
Third edition

Paul Stollard and John Abrahams

E & FN SPON

An imprint of Routledge

London and New York

First edition published 1991 by E & FN Spon, an imprint of Chapman & Hall

Second edition published 1995 by E & FN Spon, an imprint of Chapman & Hall

Third edition published 1999 by E & FN Spon, an imprint of Routledge
11 New Fetter Lane, London EC4P 4EE

Simultaneously published in the USA and Canada
by Routledge
29 West 35th Street, New York, NY 10001

Typeset in Times by Keystroke, Jacaranda Lodge, Wolverhampton
Printed and bound in Great Britain by TJ International Ltd, Padstow, Cornwall

British Library Cataloguing in Publication Data
A catalogue record for this book is available from the British Library

Library of Congress Cataloging in Publication Data
Stollard, P. (Paul), 1956–
Fire from first principles: a design guide to building fire safety/
Paul Stollard and John Abrahams. – 3rd ed.
p. cm.
Includes bibliographical references and index.
1. Building, Fireproof. 2. Safety factor in engineering.
I. Abrahams, John. II. Title.
TH1065.S69 1999
693.8'2–dc21
                                                    98–37215
                                                        CIP

ISBN 0–419–24270–8 (pbk)
ISBN 0–419–24260–0 (hbk)

Contents

Fire from First Principles

Authors

Paul Stollard Ph.D, MIFS, ARIAS, MIFireE, MIBC, BArch, BA. A Principal Architect with the Scottish Office responsible for the content of the Technical Standards. A former director of Abrahams Stollard Ltd and Visiting Professor of Architecture at the Queen's University of Belfast.

John Abrahams MIFS, ARICS, MIBC, AIFireE. A director of Abrahams Stollard Ltd, fire safety engineers and construction safety consultants.

The illustrations were prepared by Derrick Perkins.

Acknowledgement

The authors would like to thank The Interbuild Fund for their financial support in the preparation of the first edition of this book.

Introduction

Fire can be useful, but it can also be deadly. It has always fascinated and frightened; and as the proverb states, it is a good servant and a bad master. Without fire, civilization would be radically different – it might not even exist. However, the cost of fires which get out of control is high, and an average of two to three people die in fires each day in the UK.

There is a risk of fire in every building that is designed, and it is accepted that complete safety from fire is an impossible goal. The fire risks inherent in different building types are normally highlighted only when a particularly serious and fatal fire attracts public attention such as the fires in recent times at Bradford football ground and King's Cross underground station. Such major fires underline the importance of building design and remind architects of their responsibility to minimize the risks of fires in buildings.

However, fire safety is not the only objective which the architect designing a new building has to fulfil. There are a whole series of other objectives (aesthetic, functional, technological and economic) which must also be satisfied, and if the design is to be successful, these must be integrated into a coherent whole during the design process. It is the architect's responsibility to ensure that the objectives of fire safety are integrated with the more general objectives of architectural design. This book seeks to outline the fundamental principles of fire safety, so that the architects, building surveyors and others in the design team can work from first principles to ensure a safe and successful design.

Legislation attempts to set minimum standards of safety with which architects must comply; however, attempts to comply without understanding the logic behind the law will lead to either inadequate levels of safety or cumbersome compromises. The design team should not view legislation as design guidance. Legislation is written for the enforcement authorities to check that the designs being produced are intrinsically and fundamentally safe, not for the architects to use as the basis of their fire safety design.

Therefore this book does not start by describing the legislation, but begins with the first principles of fire safety. The various tactics that the architect can use to ensure the safety of the occupants and the protection of the building are outlined as the basis for design. Working from first principles and considering fire safety throughout the design process, the architect will be able to achieve both safer buildings and ones where the fire precautions are less intrusive. Fire safety measures will be far less obvious and more effective if designed in, rather than bolted on afterwards.

Although it is essential for architects to work from first principles, it is not necessary for them to become fire scientists. Therefore principles are laid out as simply and clearly as possible; and to supplement these a series of tables are included to offer approximate guidance on matters of fire escape and fire containment. These tables are intended particularly for student architects working at the sketch design stage, for whom it is more important to gain a general idea of what is required, and why, than to understand the minutiae of codes and standards. The inclusion of these rough tables does not contradict the first principles approach of the book, rather they seek to provide rules-of-thumb which designers can use to check that they are heading in at least the right general direction.

There is little value in confusing with unnecessary information, even less in presenting a spurious accuracy through complex calculations. An architect should not need to get involved in anything that requires calculation; if the building warrants this, then a fire safety consultant should probably be involved. Therefore there are no calculations or formulae within this book. Similarly, there are no references included in the text, for this is not an engineering textbook nor a treatise on fire science. The final chapter (Chapter 8) offers an annotated and structured bibliography in which further reading in specific areas is identified for those who do need to know more; and the chapter also reviews the current UK legislation and shows how this relates to the principles of fire safety.

Although primarily intended for architects, this book should also serve as a useful basic text for the statutory authorities (fire service and building control). It is even more important that these groups can understand the first principles of fire safety on which the legislation is based if they are to be able to enforce it fairly and effectively. One of the common causes of problems between the design team and the authorities is poor communications and a lack of mutual understanding. Perhaps a common textbook, working from first principles, might resolve some of these confusions?

Chapter 1 provides a brief introduction to the theory of fire safety and introduces the technical terms of fire science which the architect will come across from manufacturers and authorities. The main part of the book (Chapters 2–6) examines the five tactics available to the architect to achieve safety: prevention, communication, escape, containment and extinguishment. These are considered as design parameters and are relevant throughout the design process. Consideration of these tactics will ensure that the building not only complies with the legislation, but more important, offers a reasonable level of safety. Chapter 7 considers the fire assessment of existing buildings and, as well as outlining the basic principles, includes four examples of assessment methods. Chapter 8 summarizes the more detailed information available to the design team and outlines the basic elements of fire safety legislation in the United Kingdom. The key bodies and additional sources of advice are also listed, along with glossaries of relevant technical terms.

Theory

1

Invariably there is too little time in the design process for architects to become fully involved in the technicalities of the combustion process – and it is not really necessary. Architects do not normally want to become (and they do not need to become) fire scientists. However, to ensure a reasonable standard of fire safety without allowing it to dominate the design, it is necessary to be aware of what happens in a fire. Architects need to know what their design objectives ought to be, and how these can be achieved. Therefore this chapter briefly sketches in the few essentials that the designer should know, and places them in the context of the design process. It should not be necessary for architects to become involved in calculations, formulae and chemical symbols, and all these have been excluded here. The references in Chapter 8 provide boundless technical data, but if architects become involved in the fine detail of chemical reactions or the structure of flames, then they have gone beyond the stage where specialist advice is essential. This chapter, then, is for the 'ordinary' architect, who does not have the time or inclination to become a fire safety specialist.

1.1 Fire science

This first section is intended to provide an outline of the key stages in ignition and fire growth and to outline the principal products of combustion. Some of the most common technical terms are explained, so that the designer will be able to follow manufacturers' literature and to discuss their designs with the legislative authorities. A full glossary of fire terms is included in Chapter 8 (section 8.7), as is an index to the book (section 8.8).

1.1.1 Ignition

Combustion is a series of very rapid chemical reactions between a fuel and oxygen (usually from the air), releasing heat and light. For combustion to occur, oxygen, heat and a fuel source must all be present and the removal of any one of these three will terminate the reaction. These three ingredients of a fire are so essential that they are referred to as the **triangle of fire**. Removal of any of the three (heat, fuel or oxygen) will terminate the reaction and put out the fire.

Flames are the visible manifestation of this reaction between a gaseous fuel and oxygen. If the fuel is a gas and already mixed with the oxygen, then this is described as a **pre-mixed flame**; if the fuel is a solid or liquid and the mixing occurs only during combustion as the fuel gives off flammable vapours, then the flames are described as **diffusion flames**. The gasification of a solid or liquid fuel occurs as it is heated and chemically degrades to produce flammable volatiles. Simply heating a suitable fuel does not necessarily lead to combustion, this only occurs when the vapours given off by the fuel ignite, or are ignited.

The temperature to which a fuel has to be heated for the gases given off to flash when an **ignition source** is applied is known as the fuel's **flash point**, while the temperature to which a fuel has to be heated for the vapours given off by the fuel to sustain ignition is described as the **fire point**. If these vapours will ignite spontaneously without the application of an external flame, then it is said to have reached its **spontaneous ignition temperature**.

Therefore it is not the fuel itself which burns, but the vapours given off as the fuel is heated. Once ignition has begun and the vapours are ignited, these flames will in turn further heat the fuel and increase the rate of production of flammable vapours. For the flame to exist at the surface of the fuel, the combustion process must be self-sustaining and capable of supplying the necessary energy to maintain the flow of flammable vapours from the fuel.

In diffusion flames the rate of burning is determined by the rate of mixing of the fuel and oxygen and this is normally controlled by the degree of ventilation, the amount of fuel and the configuration of the room – all factors which the architect can influence. However, no such restrictions exist with pre-mixed flames and therefore the rate of burning can be very much faster. A common example of a pre-mixed flame is the laboratory Bunsen burner. A pre-mixture of fuel and oxygen in a confined space will lead to an explosion risk. Although a gaseous fuel can be mixed in all proportions with air, not all mixtures are flammable, and it is possible to establish the upper and lower limits of flammability outside of which a flame cannot travel through the mixture.

Throughout this section it is intended to refer to recent fire statistics, showing the practical application in buildings of the theoretical issues of fire ignition, growth and products. In 1996 there were a total of 532 500 fires attended by local authority fire services in the UK; however, less than one-third of these fires occurred in occupied buildings. In very general terms, the figures were in line with past experience and 1996 can be taken as a fairly representative year.

It is occupied buildings which are obviously of concern to the architect. In 1996 there were over 600 fatalities in occupied buildings, with the number of other injuries in excess of 16 000. It is important to distinguish between dwellings and other occupied buildings. Approximately 60% of fires in occupied buildings occurred in dwellings, but they accounted for 90% of fatalities and over 80% of non-fatal injuries (Table 1.1).

Table 1.1 Fire statistics: UK, 1996

	Fires	Fatal injuries	Non-fatal injuries
Dwellings	68 800	594	14 163
Non-dwellings	45 400	35	2 417
(All occupied buildings	114 200	629	16 580)
Refuse and derelict vehicles	162 800		
Vehicles	72 800		
Chimneys	28 700		
Grassland, etc.	117 900		
Derelict buildings	12 600		
Others	23 500		
Total	532 500		

Source: Home Office, *Summary Fire Statistics*, UK 1996 (1998), derived from tables 1, 2 and 6.

1.1.2 Fire growth

The three basic mechanisms of heat transfer are conduction, convection and radiation; and all three are common in building fires. **Conduction** is the mode of heat transfer within solids, and although it does occur in liquids and gases, it is normally masked by convection. **Convection** involves the movement of the medium and is therefore restricted to liquids and gases. **Radiation** is a form of heat transfer which does not require an intervening medium between the source and receiver (Figure 1.1).

Fires within enclosed spaces behave differently and with different rates of burning from those in the open. It is important for the architect to understand the stages in the development of an enclosed fire as they will be the most common (Figure 1.2). The presence of a 'ceiling' over a fire has an immediate effect of increasing the radiant heat returned to the surface of the fuel, and the presence of walls will also increase this effect (provided there is sufficient ventilation). With sufficient fuel and ventilation, an enclosed fire will pass through a series of stages after ignition: a period of growth, one of stability and then a period of cooling. The plotting of the temperature of a fire against time from ignition will give a **fire growth curve**, and as these will vary to reflect the conditions of the fire, they are extremely useful to fire scientists considering the consequences of changing the conditions.

The growth period lasts from the moment of ignition to the time when all combustible materials within the enclosure are alight (Figure 1.3). At first, the vapours given off by the fuel will be burning near the surface from which they are being generated; the ventilation is normally more than enough to supply oxygen for this, and the rate of burning is controlled by the surface area of the fuel. The duration of the growth period depends on many factors, but a critical moment is reached when the flames reach the ceiling. As they spread out under the ceiling, the surface area greatly increases. Consequently, the radiant

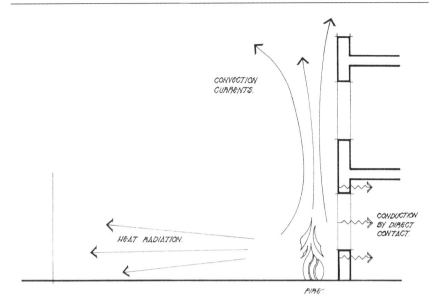

Figure 1.1 Forms of heat transfer

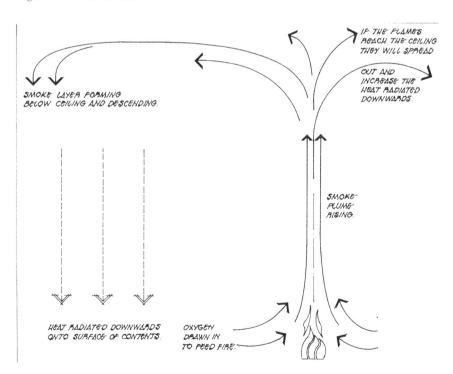

Figure 1.2 Standard compartment fire

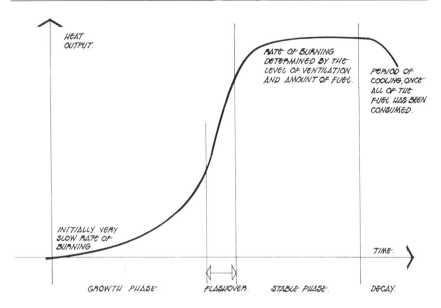

Figure 1.3 Standard fire growth curve

heat transfer back to the surface of the fuel is dramatically increased. This will probably occur (in a domestic-sized room with typical furnishings) when the temperature at the ceiling has reached approximately 550°C. The remaining combustible materials will now rapidly reach their fire points and ignite within 3–4 s. This sudden transition is known as **flashover** and represents the start of the stable phase of the fire.

If there is inadequate ventilation available to the fire during the growth period, then the fire may fail to flashover due to oxygen starvation. The fire may die out completely, or it may continue to smoulder; and such a smouldering fire can be extremely hazardous as the enclosure fills with flammable vapours. If this is then mixed with a new supply of oxygen (for example, by a door being opened), it may ignite with an eruption of flame, this effect being known as backdraught. This can be highly dangerous for firefighters attempting to enter rooms to search for survivors, and they have to ensure sealed or semi-sealed spaces are provided with some controlled ventilation at high level before attempting entry.

During the stable phase of an enclosed fire the flaming is no longer localized, but occurs throughout the enclosure. The volatiles are mixing with the incoming air, and the rate of burning will be determined by the level of ventilation and the amount of fuel present. It is this stage of the fire which is of greatest significance to the architect because maximum temperatures will be attained. The fire resistance of elements will have to take into account both the maximum temperatures which will be reached and the length of time for which they are likely to be sustained. The final period of cooling sees the decay of the fire, once all available fuel has been consumed.

Combustion can only occur if oxygen is present; many extinguishing agents operate by limiting the amount of oxygen available to the fire (e.g. carbon dioxide, foam, sand). The most common extinguishing agent (water) works by cooling the materials involved in the reaction. Without heat the reaction cannot start and if the materials are suddenly cooled the reaction will cease. The third method of extinguishing a fire is by interrupting the reaction itself, and dry powder and the halogenated hydrocarbons (commonly known as halons) do this by slowing down the reaction until it ceases to be self-sustaining.

Looking at the statistics for fires in occupied buildings, it can be seen that over one-third are confined to the first item to be ignited and more than eight out of ten do not extend beyond the room of origin. Considering dwellings and other buildings separately, it is apparent in dwellings that even fewer fires extend beyond the first item. These figures are based on the spread of fire damage, and may not fully take account of smoke spread in advance of the fire (Table 1.2).

The major products of combustion are heat, light and smoke. The smoke consists of chemicals produced by the oxidization of the fuel. They are included with fine particles of burnt and unburnt fuel, and drawn into (entrained in) a buoyant plume of heated air. Mixed with this there may also be toxic gases produced by combustion. Light is unlikely to be a danger, but the other two products (heat and smoke) are particularly dangerous and must be designed against.

1.1.3 Heat

Smoke damage to a building can be severe, but it rarely causes total collapse; however, extreme heat can completely destroy a building. Steel will have lost two-thirds of its strength by the time it has been heated to 600 °C – a by no means uncommon temperature in a domestic fire. Concrete is a more resistant material; but as reinforced concrete depends on steel for its tensile strength, there needs to be sufficient insulation of the steel to prevent it reaching its critical temperature. Timber, of course, burns but is a very good structural material as burning occurs

Table 1.2 Fire spread statistics: UK, 1992 (percentages)

	Percentage of fires in:		
	all occupied buildings	dwellings	non-dwellings
Confined to first item ignited	39	43	33
Confined to room of origin	50	48	54
Confined to building of origin	9	8	8
Spread beyond building of origin	2	1	5

Source: Home Office, *Fire Statistics: United Kingdom*, 1992 (1994), derived from table 56.

at a fairly constant rate and so structural timbers can be oversized to provide a known measure of fire resistance. Bricks provide one of the best fire-resistant materials as they have already been kiln-fired at high temperature during manufacture. The design of structural elements to resist heat is the responsibility of the architect and will be considered in more detail in the discussion of fire containment (Chapter 5).

The amount of heat produced in a fire is often regarded as the measure of the severity of the fire. An understanding of the factors which determine the level of heat production will enable an estimate to be made of the potential of the fire to destroy property, both where the fire started and in adjoining areas. In a compartment fire the rate of burning has been identified as being dependent upon the fuel and ventilation available. Therefore it is these same two factors which determine the heat which will be produced.

The quantity of the potential fuel within a building is described as that building's **fuel load**, and this will include both the fabric of the building and its contents. Estimating the fuel load can give guidance on the likely heat production and fire severity. The fuel load is difficult to establish accurately due to the multiplicity of materials likely to be involved. It is sufficient for the architect to be aware that the fuel load will vary depending upon the building contents and fabric, and this should be borne in mind during the design process. The fuel load of, say, a large distributive warehouse is likely to be much higher than the fuel load of a sports centre of similar size (Table 5.1). When considering the potential for smoke production from the same materials, the term **smoke load** is normally used. (These terms, along with technical phrases, are listed in the glossary in Chapter 8.)

It is not only the nature and amount of the fuel which will influence the heat output; the arrangement of the fuel is also significant. In essence, the greater the surface areas of fuel exposed, the greater is the potential for rapid fire development. The proximity of the fuel to the walls and ceiling will also be a factor in determining the spread along these surfaces. The denser the arrangement of the fuel, the longer the fire will take to build up to full heat output, and the longer it will last.

The ventilation of the space where a fire starts will be critical in determining the fire severity and heat output. Both the air supply to the fire and the possible loss of heat by air removal are significant. The amount of ventilation will be determined by the shape and size of the window openings. When the windows are small, the size of the fire may be limited by the amount of oxygen which can be provided (a ventilation-controlled fire). If the windows supply more oxygen than the fire needs, then the rate of burning will be controlled by fuel availability. Increasing the supply of oxygen above that which can be used in the combustion process will serve only to cool the fire as it becomes entrained into the rising smoke plume. Not only is the size of the window openings significant, their shape can also influence the fire severity. Experimental work has shown that a narrow, tall window will encourage a higher burning rate than a square window of the same area.

The final factor which has an influence on the fire severity and rate of heat output is the size of the enclosure or room in which the fire is burning. While a larger area probably contains a greater fuel load, the distance from the fire to the ceiling and the walls will slow down the fire in the initial stages. In general terms, the larger the area, the longer the fire will take to develop, but the fire will be the more severe once established.

1.1.4 Smoke

A small percentage of fire victims die indirectly due to the heat generated in a fire causing the collapse of the building, and this is a particularly acute risk for firemen and rescuers. However, the majority of fire deaths are due to smoke, either by the inhalation of toxic gases or to carbon monoxide poisoning. Various studies of fire deaths have found that over half of all the deaths were directly attributable to carbon monoxide poisoning. Very few victims are burnt to death, for although approximately two-thirds of the victims may have 'fatal' burns, almost all of these are received after death. The frequent burning or charring of bodies after death due to toxic gas or smoke inhalation can give a false impression of the relative dangers of the different fire products.

Smoke is the general term for the solid and gaseous products of the combustion in the rising plume of heated air. It contains both burnt and unburnt parts of the fuel, along with any gases given off by the chemical degradation of the fuel. The heating of the fuel and the emission of the volatiles will cause a plume of heated gases to rise, and this will entrain air at its base and as it rises. Some of this air provides the oxygen necessary to support combustion; the surplus will mix with the rising plume and becomes an inseparable element of the smoke. Although a complex phenomenon, smoke can be treated by architects as a single problem because they will be designing against the mixture rather than individual constituents. The architect must consider all smoke as dangerous and attempt to limit its production and control its movement.

By far the largest of the constituents of smoke is the air which is entrained, and therefore in any attempt to estimate the rate of smoke production it is sufficient to assess the rate of air entrainment. This clearly depends on the size of the fire (in particular, its perimeter and the height of the rising smoke column) and the intensity of the fire (in particular, its heat output). In most buildings it is impossible to calculate accurately the rate of smoke production because of the large number of variables, and it is sufficient for the architect to realize that the larger the fire (and the larger its perimeter), then the greater the rate of smoke production. Sprinkler systems are normally designed to limit a fire to a 9 m^2 area (a 12 m perimeter), and therefore in calculating smoke production from fires in sprinklered buildings, it is assumed that this represents the largest probable fire area.

The appearance of smoke reflects its constituents and will vary from a very light colour to a deep sooty black. The density of smoke depends on the amount of unburnt particles carried in the air, and the more dense is the smoke, the more dangerous it is because of reduction in visibility. Visibility in smoke depends both on the smoke density and on the psychological condition of the observer. A very dilute smoke may only be an inconvenience, but when visibility is seriously curtailed, it can prevent escape and hence be extremely dangerous. Diluting smoke sufficiently to keep escape routes open is almost impossible, so it is naturally better to prevent smoke entering in the first place.

The architect must consider all smoke as potentially lethal, though the toxicity will vary depending on the nature of the fuel. All carbon-based materials will give off carbon dioxide and carbon monoxide. Even more toxic gases will result from other fuels, and hydrogen chloride, hydrogen cyanide and the oxides of nitrogen are all common in fires. It is thought that combinations of these gases are even more toxic than when occurring individually, due to the effect known as synergy, but the chemistry of such reactions is too complex and too little understood to be relevant to designers. It is more important for architects to be aware of the particular dangers of certain materials. Polyurethane foam, in particular, produces large amounts of hydrogen cyanide which is lethal in very small quantities. The 1979 fire at the Woolworth's store in Manchester was a graphic example of the lethal nature of the smoke produced by burning polyurethane. Most of those who died were within feet of the escape stairs, but could not to reach them because of the very rapid toxic smoke spread.

Considering the injury statistics of fires it can be seen that the pattern in dwellings is quite different from that in other building types. In dwellings nearly nine out of ten injuries are the result of accidental fires, with the most common causes being smokers' materials (over one-third of fatalities, and one-sixth of non-fatal injuries), space heating appliances (one-eighth of fatalities, and one-twelfth of non-fatal injuries) and matches (approximately one-twelfth of all injuries). Cooking appliances also cause a large number (over one-third) of non-fatal injuries, but relatively fewer fatalities. In occupied buildings other than dwellings over one-third of all injuries resulted from deliberate fires. While among accidental fires, the most likely to lead to injuries were smoking materials and electrical appliances. There were also a large number of deaths and injuries attributable to 'other sources', mainly connected with industrial or constructional processes (Table 1.3).

1.2 Fire safety design

The previous sections have highlighted the dangers of fire in terms of the products of combustion, namely heat and smoke. The objectives of the architect in designing buildings which offer an acceptable level of fire safety are concerned with minimizing the risks from these products; the architect must be conscious of these objectives, and of the tactics available to achieve them.

The precise measures taken to achieve the fulfilment of those objectives through specific tactics can be considered as the components of fire safety. And

Table 1.3 Injuries, by source of ignition statistics: UK, 1992 (percentages)

	Percentage of injuries in: all occupied buildings		dwellings		non-dwellings	
	fatal	*non-fatal*	*fatal*	*non-fatal*	*fatal*	*non-fatal*
Smokers' materials	35	16	36	17	23	10
Matches	6	6	7	7	3	5
Cooking appliances	9	33	10	38	5	7
Space heating appliances	12	7	13	7	6	4
Electrical wiring	1	4	1	3	1	5
Other electrical appliances	4	8	5	7	0	9
Others (including natural phenomena)	7	7	6	4	13	22
Unknown	13	2	11	2	22	3
(All accidental	87	83	89	86	73	65)
Deliberate fire-raising	13	17	11	14	27	35

Source: As Table 1.2, derived from tables 6, 10, 11, and 31.

these components concern what is actually built or installed – the fire doors, sprinklers, escape stairs, etc. It is essential not to confuse such specific components with the more general tactics and objectives which the architects must also follow. Compartmentation is a valuable tool to the designer, but used without understanding, it does not constitute an effective tactic for fire containment or achieve designated objectives. In order to achieve this, it is essential that the designer has an understanding of the principles underlying fire safety (Figure 1.4).

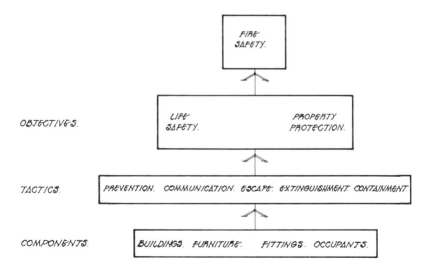

Figure 1.4 Objectives, tactics, components hierarchy

1.2.1 Fire safety objectives

The design process can be viewed as the attempt by the architect to satisfy a series of objectives: it is the search for a physical solution to a given set of problems. These objectives will include aesthetic, functional, technological and economic issues. If the building is to be successful, then the objectives will need to be integrated into a coherent and balanced whole. Among the technological objectives will be those of fire safety.

Fire safety is normally considered to cover both the safety of people and of property – in the building concerned and in the surrounding area. Therefore the **fire safety objectives** for the architect will be twofold: life safety and property protection. Other objectives might sometimes be relevant, but these will normally only be variations on, or combinations of, those two principles. For example, in the fire safety design of hospitals the maintenance of the service is considered an objective (to avoid consequential loss of life due to postponed operations and treatment), yet this is only a variation on life safety and property protection rather than a completely new objective.

In designing to ensure life safety the architect is seeking to reduce to within acceptable limits the potential for injury or death to the occupants of the building and for others who may become involved. The objective of property protection is the reduction to acceptable limits of the potential for damage to the building fabric and contents. The architect will be seeking to ensure that as much as possible of the building can continue to function after a fire, and that the building can be repaired. The building should also remain safe for firefighting operations during the fire. The risk to adjoining properties will have to be considered, as well as the wider hazard of possible environmental pollution.

The two principal products of combustion relate to those two objectives and, in very crude terms, life safety can be seen as protecting people from smoke, while property protection concerns keeping heat away from the building. This gross oversimplification provides a succinct summary of the objectives which the architects must fulfil and the dangers they must avoid.

1.2.2 Fire safety tactics

There are five **tactics** available to the architect seeking to fulfil the objectives of life safety and property protection (Figure 1.5).

1. **Prevention** – ensuring that fires do not start by controlling ignition and fuel sources.
2. **Communication** – ensuring that if ignition occurs, the occupants are informed and any active fire systems are triggered.
3. **Escape** – ensuring that the occupants of the building and the surrounding areas are able to move to places of safety before they are threatened by the heat and smoke.

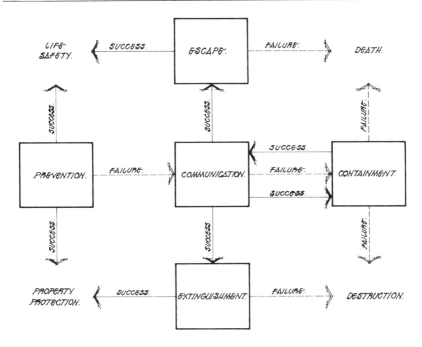

Figure 1.5 Matrix of tactics and objectives

4. **Containment** – ensuring that the fire is contained to the smallest possible area limiting the amount of property likely to be damaged and the threat to life safety.
5. **Extinguishment** – ensuring that the fire can be extinguished quickly and with minimum consequential damage to the building.

Considering the five tactics in a logical sequence, the first is obviously prevention and only if this fails are the other tactics attempted. If fire prevention is successful, the others need not be attempted; however, as fire prevention will inevitably fail at some stage during the life of the building, provision must be made for the other tactics.

Communication by itself, even if totally successful, cannot save lives or protect property, but its key role in ensuring fire safety means that it must be considered as one of the five tactics. If communication is successful, then escape and extinguishment can be attempted; but if it is unsuccessful, then only containment remains as an available tactic.

The above five tactics provide the fundamental framework within which the architect should be working. A building designed with adequate consideration given to these five factors will offer an acceptable level of fire safety. Each of the tactics will be the subject of one of the chapters in this book, with consideration of the implementation of the tactic through the design process from inception to completion. The relationship between these fundamental fire safety tactics and the current legislation which architects have to comply with is an interesting one, and this will be considered later under the section on acceptability and equivalency.

It is interesting to note that in the British Standard Institution's proposals for the replacement of BS 5588 (Code of Practice for Fire Precautions in the Design of Buildings), the new BS 9999 will have four 'framework' documents which reflect the five tactics. It is proposed that these documents will be: Means of Escape (Escape), Construction (Containment), Firefighting Facilities (Extinguishment), and Fire Safety Management (Prevention and Communication).

1.2.3 Fire safety components

The **fire safety components** are the weapons that the designer can use tactically to achieve fire safety. They are the building itself, its furniture, fittings and occupants. The number of components is limitless and depends solely on how they are categorized. It is not just the obvious (such as fire extinguishers) which must be included, everything from the wall coverings to the management of the occupants may be relevant.

Each of the components may contribute to one or all of the five tactics, and it is this complexity of interaction which necessitates a logical approach to the tactics of fire safety. There will also be interactions between the objectives, the tactics and between any of the individual components. For this reason, measures taken to reduce the fire risk or hazard cannot be viewed in isolation, and the over-all impact of any measures must be considered.

For instance, the provision of sprinklers in a building to improve the property protection may reduce the risk of a fire growing beyond certain limits. This restriction in fire size and rate of subsequent fire growth should reduce the risk of structural failure and limit the amount of smoke produced. It should also increase the amount of time available for escape by containment of the fire. However, it will probably also reduce the smoke temperature and this will increase the possibility of local smoke logging. These problems of smoke control may result in an increased risk of life loss. There is also the risk that the sprinklers may not function efficiently and this may alter the risks both to life and to property.

The fire safety decisions are therefore complex ones, and the designer has to be aware that changing one component or altering the emphasis placed on one of the five tactics can have an effect on the probability of success in each of the other objectives. In the more complex designs it may be necessary to attempt to consider such interactions quantitatively, but for most projects it is sufficient for the designer to be aware of the possible implications of his or her decisions.

1.2.4 Acceptability and equivalency

Absolute safety from fire, where there is no risk whatsoever, is an ideal which it is impossible ever to achieve. The architect is never asked to provide such absolute safety, only to reduce the risks to property and people to a level which society regards as acceptable.

This **acceptable level of safety** has traditionally been defined through legislation. However, legislation tends to be produced as a response to particular problems or fires and it does not always offer a balanced or reasoned structure for fire safety. There is an argument that the whole history of fire safety legislation is simply a catalogue of responses to serious fires. It can be shown that not only is legislation enacted in response to disaster, but also that changes in building forms and technologies normally occur in response to disaster. In the case of the Bradford City football ground (1985) fire, the Home Secretary had announced within two days that higher safety standards were to be required at third- and fourth-division football grounds, yet this was the first fire at a football ground in which a member of the public had been killed. Public reaction, however, demanded such moves and within two days of the fire six other football clubs had either closed stands or started to remove perimeter fences and provide exit gates. But these reactions were not based on any rigorous assessment of risk, and the legislation did not form part of any general or comprehensive strategy for a common level of safety in all buildings. The vast majority of people who die in fires die in their own homes (Table 1.1), yet because these are small incidents which rarely gain media coverage the standards of fire safety required in domestic dwellings are possibly lower than in most other buildings.

With much of the legislation related to fire safety being introduced responsively following particular tragedies, the existence of a coherent fire safety policy can be queried. Society is happy to accept as safe all buildings in which the dangers have not recently been exposed by a serious fire. Compliance with the regulations which are in force at one particular time is assumed to provide an acceptable level of safety, even though that level cannot be objectively quantified.

There are alternative ways of measuring safety other than showing compliance with legislative standards. It is possible to design against specified risk criteria, for example, designing to ensure that the probability is of a single death occurring once every thousand years and a multiple death once every million years. It is also possible to design against specified probabilistic or deterministic criteria (i.e. to ensure that people within a building are able to reach a place of safety in a time less than that for conditions within the building to become untenable).

It is obvious that there is a point beyond which any increase in fire protection measures adds to the cost in undue proportion to the added safety provided. The fire safety designer must therefore achieve a balance between safety, economics and convenience. Acceptability must be discussed in view of the fact that absolute safety cannot be achieved, and of the law of diminishing returns (i.e. a type of cost–benefit analysis).

Another important concept, linked closely to acceptability, is **equivalency**. Once architects are able to achieve an agreed acceptable level of safety by whatever combination of tactics they chose, then the importance of ensuring equivalence is critical. Equivalence between two different fire safety designs means that they achieve the same level of safety by different methods. This is sometimes described as **'trade-off'**, the concept that one safety measure is being traded-off for another; for example, does a concentration on fire escape enable the architect to pay less attention to fire extinguishment, or do measures installed to decrease the possibility of ignition balance a decrease in fire containment measures?

Attempts to assess equivalency in terms of a single numerical value are difficult and can hide a number of contradictions. For example, one might regard the presence of sprinklers as providing an additional level of safety which would permit a reduced escape distance. This would then perhaps reduce the annual loss of life with an occasional larger loss, resulting from the one in 50 times that the sprinklers fail. A strategy for equivalency must recognize the distinction between average and societal risk. Calculations of equivalence are therefore neither simple nor easy to quantify; and for this reason, they are not normally explicitly incorporated into the legislative framework.

Many of the current Building Regulations are couched in terms of what is 'adequate' or 'reasonable'; however, these terms are not defined except by reference to approved documents which prescribe how each component must be designed. There is no attempt to define either acceptable safety or a system of determining equivalency, yet by saying that the approved documents constitute what is 'adequate' and 'reasonable', then any alternative fire safety strategy offering equivalent or better levels of safety should be acceptable. If designers are to be able to satisfy the objectives of fire safety without compromising other objectives (economics, aesthetic considerations and functionality), they need to be aware of the full range of different (but equivalent) fire safety strategies.

1.2.5 Traditional and fire engineering approaches to fire safety design

The traditional approach to fire safety as espoused through the Approved Documents of the Building Regulations, has been to identify certain components and then to prescribe certain standards for these components. Such components in the current Building Regulations for England and Wales which apply to new buildings include:

- travel distances and routes;
- loadbearing elements of the structure;
- roof construction;
- separating walls;
- compartment walls and compartment floors;

- protected shafts;
- concealed spaces and fire stopping;
- internal surfaces;
- stairways.

The Fire Precautions Act 1971 which covers existing premises, as well as new developments in certain building types, also covers in addition such issues as:

- staff training;
- fire brigade access;
- manual firefighting equipment;
- detection and alarm systems;
- emergency signs and lighting.

Such an incremental approach has already been criticized for limiting the design choice of architects, giving no guidance on acceptability and not helping in any calculation of equivalency. However, the most serious criticism of such an approach is the total neglect of some aspects of fire safety; fire prevention is hardly mentioned and smoke control is not given the attention it warrants. The traditional approach was to regard all these components of fire safety as somehow independent and to demand a 'reasonable' standard of provision in each of them. This limits the design flexibility of the architect and can lead to resentment of the legislation; and the architect starts to seek loopholes or ways to get round the legislation. Such an approach guarantees that the designers and the legislative authorities will be in opposition to each other.

The traditional approach also creates an artificial distinction between the requirements of the legislation which are normally supposed to concentrate on life safety, and those of the building's insurers which will be more concerned with property protection. Yet most fire safety measures will contribute to some degree to both life safety and property protection. The artificial separation of the two can lead to examples both of areas of overlap and gaps. Conflict can even be generated by the differing priorities of the legislation and the insurers: in one Oxford shopping centre the insurers offered a reduced premium if the mall was sprinklered, but this was not done because of the increased risk to life that the architects felt would result.

The alternative approach to fire safety can be described as the fire engineering approach, and it is the one underlying this book. In this, the building is considered as a complex system, with the fire safety design just one of the many interrelated subsystems. The architect is faced with designing not just to satisfy a series of prescriptive standards, but to achieve an acceptable level of safety. This will require an assessment of the equivalence of alternative fire safety strategies and the development of an integrated approach to fire safety. In the fire engineering approach, issues like fire avoidance and fire communication can be given their due weight and the designer can fully exploit all the techniques of improving fire safety. The fire engineering approach demands an understanding by the designer of the fundamentals of fire safety, but it offers the opportunity to attempt unconventional ways of achieving compliance with the legislation. All that designers are required to do is prove that what they are offering represents a level of safety equivalent to what has been defined as the acceptable level.

The fire engineering approach will begin with the recognition of the objectives of fire safety, life safety and property protection. Then the fire safety design must be prepared on the basis of an assessment risk and an analysis of the possibilities for protection by fire safety measures. Chapter 7 discusses the principles of fire assessment, and the specific risks are considered in more detail in Chapter 2 on fire prevention. There must also be a corresponding assessment of the safety offered through the design, including the potential for communications, escape, containment and extinguishment. These fire safety tactics will be considered respectively in Chapters 3–6.

Such assessments of risks and precautions may be either qualitative or quantitative. Qualitative techniques rely on an expert-based assessment of risk. Quantitative techniques are also available (e.g. fire growth modelling, smoke models, structural models, probabilistic analysis, structural response modelling, environmental testing and extrapolation of results, deterministic calculations, stochastic evaluation, fault tree, event tree and critical path analysis). However, at present such quantitative models are too complex in form and too limited in scope to be of significant value to the 'ordinary' architect. It is hoped that more accessible quantitative models will be developed which can be used easily by designers to enable them to assess the fire safety of their proposals. Chapter 7 includes four examples of fire assessment methods, for hospitals, dwellings, workplaces and canal tunnels.

This book is structured around the fire engineering approach to fire safety and each of the following chapters addresses one of the five tactics available to the architect in preparing a fire safety design. Although, sadly, it is not yet possible to offer a simple quantitative system which would permit a full study of the equivalency of alternative proposals, an understanding of the fundamentals will enable the architect to make approximate judgements on equivalency and on the acceptability of the proposed fire safety measures.

The British Standards Institution has published a Draft for Development (DD240) proposing a framework for quantitative fire safety engineering, and this is starting to be used on major projects in the United Kingdom. It is based on a series of subsystems which enable the relationship between a developing fire and the design of a building to be examined in a 'qualitative design review' and a quantitative analysis of the design against given criteria. The major limitation on the use of DD240 lies in its inevitable complexity. It does not yet offer a fire safety engineering framework of sufficient simplicity and reliability for its use to become widespread in all building types.

For most designers and architects the hope must be that the gradual improvement of the Building Regulations will lead to a stage when the safety demanded by legislation will be coherent and balanced for every building type. Designers should be asked to achieve levels of safety defined in terms of risk to people and property rather than in the prescriptive standards of compartment sizes, door widths or travel distances. Throughout this book, and in any approach to fire safety based on first principles, architects must consider safety in this manner.

Prevention

2

The simplest and most effective tactic available to the architect to ensure fire safety is to prevent fires starting, namely fire prevention. If this tactic is successful, then there is no need even to attempt any other fire safety measure. There are two ways of preventing fires and they are related to the fundamental 'triangle of fire', outlined in Chapter 1. The three elements of the triangle are an ignition source, a fuel and a supply of oxygen, and as it is almost impossible (and most undesirable) to exclude the oxygen from a habitable building, fire prevention has to concentrate on the other two elements. Prevention of ignition and the limitation of the fuel available are the twin methods of fire prevention. There is also a minor role for the architect in ensuring that the plans for the fire safety management of the building are properly prepared, and this will be considered at the end of this chapter.

2.1 Ignition prevention

In designing to reduce the **ignition risk** the architect has to do two things: first, to design out the predicted ignition hazards or sources; and secondly, to enable the building to be managed in such a way that the risk of ignition is eliminated. The actual design against the risk and the design to permit management against the risk must be seen together.

The first necessity for the designer is an understanding of the most likely ignition risks in the particular building type under consideration: it is essential to know your enemy if it is going to be defeated. There are four main classes of ignition:

1. natural phenomena (e.g. lightning);
2. human carelessness (e.g. smoking materials, matches, cooking);
3. technological failure (e.g. electrical wiring and appliance faults);
4. deliberate fire-raising (e.g. suicide, vandalism).

These four categories are not mutually exclusive, and technological failure, in particular, is normally in part the result of human carelessness. Technology alone cannot be held to be the culprit when it is the human misuse of that technology which has caused the problem.

Considering the fire statistics on sources of ignition for occupied buildings in the UK, in 1996, a significant difference can be detected between dwellings and other occupied buildings. In dwellings almost one fire in four was deliberate, and among those which were accidental the most common sources of ignition were cooking appliances (nearly three-fifths of all fires) and smokers' materials (one-twelfth). In non-domestic buildings the percentage of deliberate fires was much higher, and among accidental sources the most common were electrical appliances. There were also many 'other' accidental sources in non-domestic buildings, reflecting the high number of fires caused by industrial or constructional processes.

It has to be remembered that it is not just the number of fires which is significant, their severity is also most important. The figures on fire casualties given in Chapter 1 show that although certain sources may be very common, they do not lead to an equivalently high level of casualties. For example, cooking appliances account for a high proportion of fires in dwellings, but not nearly as many fatalities as do smokers' materials. This reinforces the importance of an appreciation not only of what is likely to cause a fire, but also which fires are the most dangerous (Table 2.1).

2.1.1 Natural phenomena

The most serious natural ignition source is lightning, and the dangers of a lightning strike are well known; the 1984 fire at York Minster highlighted this in the most dramatic manner. Earthquakes are also major fire risks through the

Table 2.1 Sources of ignition statistics: UK, 1996 (percentages)

	Percentage of fires in: all occupied buildings	*dwellings*	*non-dwellings*
Smokers' materials	7	8	6
Matches	1	1	1
Cooking appliances	28	41	10
Space heating appliances	4	5	3
Electrical wiring	4	3	4
Other electrical appliances	11	10	13
Others (including natural phenomena)	10	8	12
Unknown	2	1	3
(All accidental	67	77	52)
Deliberate fire-raising	23	23	48

Source: Home Office, *Summary Fire Statistics*, UK, 1996 (1998), derived from tables 2 and 3.

damage they can cause to gas and electricity supplies, and fire is the consequent problem in earthquake areas. Forest fires are an additional natural hazard for buildings adjoining or surrounded by large areas of woodland. In extreme situations, buildings may even be threatened by volcanic activity. However, in the UK the only major concern for architects is lightning, and all architects should be aware of how to design against this hazard.

The average lightning flash normally lasts for less than a thousandth of a second, but during this time a vast quantity of electrical energy is dissipated to the earth (perhaps a current of 10 000–100 000 A and a voltage of many millions of volts). The lightning stroke may even be repeated over the same path two or three times within a couple of seconds. Lightning damages buildings as the current passes through building materials or along crevices between them, and energy is dissipated with the heat reacting to the water content of the materials to produce very hot gases.

The buildings most at risk are those at high altitudes, on hilltops or hillsides, in isolated positions, and of course those with tall towers or chimneys. Designers must ensure that buildings at risk are provided with lightning-conductor systems which will dissipate the shock directly to the ground. The finials of the lightning conductors need to be mounted on the highest points of the building and they have to be linked to the ground by a 'down conductor', normally of copper tape. This should be attached to the outside face of the building, and at ground level the down conductor must be linked to an earth termination (normally a copper rod driven some 3 m into the ground, or a copper plate buried below the surface).

Experience shows that a lightning conductor on a tall building will protect everything within a cone extending around the building for a distance equivalent to the height of the conductor (sometimes described as 'bell-tent protection'). Buildings, or parts of buildings, outside or reaching above this cone will need their own protection. It has to be stressed that lightning is one of the most complex of natural phenomena and some of its manifestations are still far from understood. It is ironic that it was York Minster which suffered so badly when it had, at the time, perhaps one of the best lightning protection systems of any cathedral in the country.

2.1.2 Human carelessness

Probably the most common cause of ignition, and certainly the hardest to design against, is human carelessness. Almost all fires started by smoking materials or matches could be avoided, and yet these are one of the major causes of domestic fires and consequent loss of life. Similarly, the continuing high incidence of fires concerned with cookers and stoves (in particular, chip-pans) are normally due to human carelessness. Issues of public education and the encouragement of good home safety are beyond the scope of this book, but there are some contributions which the designer can make – the simplest of all being not to site the cooker master switch behind the cooker, but to one side and in an accessible position.

In non-domestic buildings there is even more that the architect can do in, most significantly, the provision of adequate storage space. Insufficient or badly sited storage will inevitably mean that the workers within a building will start to store goods in corridors, kitchens or wherever it is easiest for them. This may well bring potential fuel materials into contact with potential ignition sources. For example, in hospitals it is highly dangerous if staff start to store flammable materials (such as linen or disposable bedpans) in areas with an ignition hazard (such as kitchens or treatment rooms).

With smoking posing such a high ignition hazard, it is also important that the architect and the client seriously consider the areas of any building where staff may go to smoke, both officially designated rest areas and the unofficial places for a quiet cigarette. A simple 'no smoking' policy by an employer will not solve the problem and may, in fact, increase the dangers if staff are forced to smoke surreptitiously in less frequented areas. This is a particular problem in warehouses and storage areas.

It is essential that in designing the layout of the building the architect has an awareness of how it will be used in practice and makes at least a guess at how it may be misused. Areas with a high ignition hazard (e.g. restrooms and kitchens) should be placed as far as convenient from areas with a high life or property risk (e.g. bedrooms or special storerooms). This simple method of ignition prevention can even be applied to domestic dwellings where the biggest ignition hazards are in the kitchen (cooking) and the living-room (smoking) and the biggest life risks are in the bedrooms. The architect should ensure that the escape routes from bedrooms are not jeopardized by passing through the living or kitchen areas. The siting of rubbish areas and skips is equally important as they are very common sources of ignition (Figure 2.1).

2.1.3 Technological failure

Ignitions resulting from technological failure are even more the responsibility of the designer and need to be consciously considered during the planning process. Just as perfect fire safety is unobtainable, so it is inevitable that all building services and systems will eventually fail. The architect must design such that failure is predictable, controllable and repairable.

In the layout of the building the architect must be aware of the areas which pose the greatest hazard in the event of technological failure and design to minimize the consequences that would result from such ignition. Areas such as plant-rooms, laboratories, boiler houses and large kitchens need to be sited where their threat is minimized. The separation within a building of those areas where there is an increased threat, and those areas where there is an increased risk to life or property, is important. For example, in a large factory the paint spray shop would be kept at some distance from the main stores, and in a shopping centre the pedestrian malls should be isolated from the rubbish and plant areas.

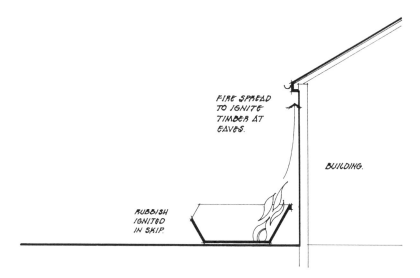

Figure 2.1 Human carelessness – skip fires

The services within a building (notably the electrical installation) will always be a major ignition hazard, and the architect must tackle this in both the short- and long-term life of the building. In the short term it must be ensured that the services and installations are correctly designed, specified, constructed, checked and commissioned. In the long term it must be ensured that the maintenance manual for the building outlines checking and replacement cycles for the installation, so that safety standards can be maintained.

The maintenance manual for a building is a crucial fire safety document specifying precisely what is necessary to ensure that all the services are maintained to the necessary standards. It is in architects' own interests to see that this document is as comprehensive as possible because it will relieve them of some of their liability for the building, transferring it to the building's owners and occupants. The maintenance manual should cover all services (electricity, gas, communications and water), the lifts and the building's active fire safety systems (alarms, detections, smoke control, auto-suppression, etc.). The maintenance manual is also a sensible place to record particular materials or building elements which require special attention because of their role in the fire safety of the building. It may be that fire-resisting or fire-retardant materials have been used, but these may not be immediately obvious to the building's occupiers (for example, fire-resistant glass, intumescent coatings or fire-retardant paints). Such materials will require particular care and should not, of course, be repaired or replaced by ordinary materials.

2.1.4 Deliberate fire-raising

It is often very difficult to prove to the satisfaction of a court of law that a fire was deliberately started, and many fires which were probably deliberate do not appear as such in the statistics. There are five main categories of deliberate fires, some easier to design against than others. They may be started for financial gain or to conceal a crime, or as malicious vandalism, casual vandalism or as terrorist acts.

Fires for financial gain include those lit by the owners or occupiers – the classic insurance frauds or attempts to resolve a company's financial problems by destroying plant or buildings. Alternatively, the fire may be started by an outsider or competitor who stands to gain from the destruction of the buildings or company. As such fires are likely to be carefully planned, it is almost impossible for the architect to design against them; anyone who is determined to burn down a building will eventually succeed, especially if they already have an intimate knowledge of the layout and construction. Fire-raisers are likely to want to conceal their crime as an accident, therefore the only defence the architect can offer is to eliminate the opportunities for accidental fires (as in the earlier section).

Again, there is little the architect can do in the design to prevent fires which are started to conceal a crime. Someone who wishes to hide the traces of a murder or a burglary by burning down the building will not be deterred. However, it is most likely that he or she will be caught as unplanned fires very rarely succeed in destroying all the evidence.

Unfortunately, fires started by acts of malicious vandalism are not uncommon and the classic example is of individuals or groups who attempt to gain revenge by starting fires, perhaps the dismissed worker who, rightly or wrongly, has a grievance against an employer and decides to retaliate by destroying the factory, shop or office. As with fires ignited for financial gain, it is probable that such a fire-raiser will have a good knowledge of the building and the working practices, so there is little the designer can do to prevent such fires.

The statistics show that a growing number of fires in the UK which appear to have been ignited as a result of casual vandalism have grown to cause major damage. These are distinct from acts of malicious vandalism because there is less intent to destroy, and less planning of the fire. This is particularly true of the recent series of fires in schools, where the entire school has been destroyed, although it is not certain that this was the prime objective of those who started the fire. The architect can do much to reduce the risk of such fires by controlling access to the building or to particular areas of the building.

There are three lines of defence (Figure 2.2) around a building: first, the perimeter of the site; second, the building face; and third, the divisions between different parts within the building. Obviously the amount of damage a fire causes will increase as each of these lines of defence is breached. At the first line of defence, it is desirable to have some form of fence, hedge or other barrier, and

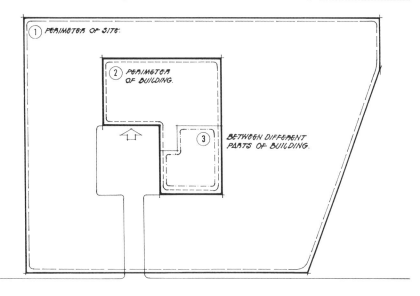

Figure 2.2 Three lines of defence

permitted entrances should be marked by at least symbolic gateways. Good lighting can also play an important role in reducing the risk of unauthorized access. At the building face the architect wants to control the number of entrances and ensure that the grounds of the building are protected by passive surveillance. Such surveillance does not necessitate continuous observation, rather it should convey to would-be intruders the feeling that they are indeed being observed. Obviously particular fire risks should not be left exposed at the building face, whether these are rubbish areas or simple storerooms. The fire at the Summerland complex on the Isle of Man (1973) was caused by young boys smoking in a small shed, which when it accidentally ignited collapsed against the outer face of the leisure centre, in turn igniting this and eventually causing 50 deaths. The third line of defence is within the building, and here the consideration of circulation routes is very important. Normally non-staff circulation should be kept to a minimum, and where large public spaces are inevitable, staff circulation should be planned to provide passive surveillance. Closed-circuit television now offers an additional means of extending surveillance and, again, it is not necessary to have someone always monitoring the system for it to be effective in deterring intruders.

The final category of deliberate ignition is terrorist attack. Fortunately, this is still a rare occurrence in most of the UK but the designers of buildings which could be the target of terrorist attack should certainly be aware of this risk. Not only are government and military buildings at risk, for recent campaigns by groups such as the Animal Liberation Front have included retail outlets and university buildings. An architect involved on a sensitive building should seek to protect against both incendiary attack and the risk of fire consequent upon the detonation of high-explosives. While the technical details of anti-explosive design are beyond the scope of this book, the greatest impact the architect can have is probably in the control of access to the building, and here the measures are identical to those we have already considered in the prevention of vandalism.

2.2 Fuel limitation

Fuel limitation, like ignition prevention, is determined by the success both of design and management measures. It is most definitely an area where the architect can play a significant role; though unless the building is managed and used as intended by the briefing and design team, it will be impossible for fire prevention measures to work most efficiently. It is hard to separate design from management and, for this reason, they will be considered together.

Limitation of the amount of fuel available will help to reduce the dangers of fire in two ways. First, by controlling the amount of material which will be able to burn and release heat to feed the growth of the fire. This is described as the **fire load** of the fuel. Second, it will control the amount of smoke which can be produced. The amount of potential fuel which will burn to produce smoke is often described as the smoke load, and this may be different from the fuel load, depending on the smoke-generating characteristics of the material involved. It is possible for a fuel to have a low smoke load and a high fire load, or vice versa. Two types of fuel are under the influence of the architect: the building fabric and the building's contents.

2.2.1 Building fabric

One of the problems for the architect is the plethora of terms used to describe the fire safety of materials. Unfortunately, it is not always possible to define a material as safe or unsafe, for it is necessary to know a little more about the conditions under which it is safe. Architects have to be wary of manufacturers who simply assure them that a material has a fire certificate, it is essential for the designer to know which certificate and what that means. The confusion which can result from the failure to comprehend the differences in fire safety terms can be clearly seen in such terms as 'ignitability' and 'fire propagation': although materials might be treated to make them hard to ignite from a small source (ignitability), such treatments may not affect their rate of burning (fire propagation), once ignited.

The essential characteristics of building materials which can be measured, (Figure 2.3) and which the architect should be aware of, are as follows.

1. **Ignitability** – the ease with which a material can be ignited when subjected to a flame.

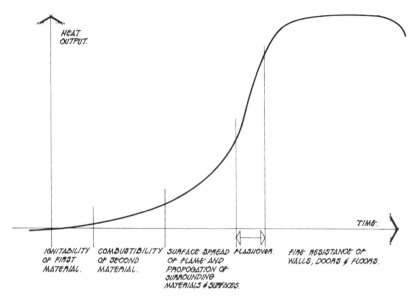

Figure 2.3 Building materials and fire growth

2. **Combustibility** – whether or not a material will burn when subjected to heat from an already existing fire.
3. **Fire propagation** – the degree to which a material will contribute to the spread of a fire through heat release when it is itself heated; this is concerned with the level of heat emission and the rate of heat release.
4. **Surface spread of flame** – whether a material will support the spread of flame across its surfaces.
5. Potential for **smoke obscuration** – the degree to which the material when burning will produce smoke that leads to reduced visibility.
6. **Fire resistance** – whether or not a component, or assembly of components, will resist fire by retaining their loadbearing capacity, integrity and their insulating properties.

There are tests and standards for each of these properties, and these are outlined in Chapter 8. The first property, ignitability, most definitely concerns ignition prevention. The next three properties (combustibility, fire propagation and surface spread of flame) are concerned with fuel limitation as they will determine how fast a fire will spread in the early stages of fire growth. The fifth property (potential for smoke obscuration) is of value because it serves to indicate the smoke load of the materials. The final property (fire resistance) should not be confused with the preceding five as it concerns the ability of a building component or assembly (rather than a material) to resist the spread of fire and, as such, it is an aspect of fire containment and will be considered in more detail in Chapter 5.

The structural elements of the building (walls, floors, roofs, ceilings, beams, etc.) should never be potential fuel sources because they must remain in place both for structural stability and to contain the fire. Where the structural elements do become fuel for the fire, as for instance with the Oroglass used in the Summerland leisure centre (1973), a major disaster can occur. The fire resistance of structural elements will be considered in Chapter 5 on fire containment.

The interior finishes on walls and ceilings are more likely fuel sources and need to be carefully selected by the specifier. If they are heated by a small fire, the linings can (because of their large surface areas) rapidly spread the fire by the radiant heating of more distant materials. Good finishes include:

- brickwork
- blockwork
- concrete
- plasterboard
- ceramic tiles
- glazing
- plaster finishes
- woodwool slabs
- vinyl wallpapers

Finishes to avoid or treat with caution include:

- timber
- hardboard
- particleboard (chipboard)
- plastics
- decorative laminates
- polystyrene wall and ceiling linings
- heavy flock wallpapers
- fibreboard

It is possible to use flame-retardant treatments to improve the safety of some of these materials, either by surface application or impregnation. The suitability of such treatments is always dependent upon their durability and their proper application.

The fire properties of any interior finish are influenced by the materials behind the surface finish, and it is essential that the architect not only specifies the finish, but also the substrate with fire safety in mind. This is a particular problem in the upgrading of existing buildings, where it is not always easy to remove existing finishes. The King's Cross fire (1987) provided a graphic example of the dangers of regularly adding new coats of paint over an extended period of time.

Certain plastics can be extremely difficult to test because they soften under heat and may even melt. If this happens before ignition, then they will not contribute significantly to fire spread, provided that they fall away from the fire. However, materials which ignite before falling can encourage very rapid fire spread. The performance of plastics may depend upon the fixings and the sheet thickness, as well as the nature of the plastic.

2.2.2 Building contents

A high proportion of fires start by the ignition of the building's contents. So where the provision of textiles, furnishings or furniture is under the control of the design team, it is important that their contribution to fire prevention is considered.

There is a separate set of terms and tests relating to fabrics and furnishings and these may be confusing both to architects and specifiers. The ignition sources used in tests are numbered, with a smouldering cigarette being source 0 and a lighted match source 1, and so on. In buildings where people sleep, or where large numbers gather, upholstered chairs should be to the standard of source 5.

Hazards arising from the burning of furniture, furnishings and fittings depend upon their construction and, in particular, the types of padding used. There is no truly 'flameproof' material and the architect or specifier can only attempt to minimize the hazard by careful choice of furnishing fabrics and foams. The particular dangers of polyurethane foam have already been mentioned in the context of the Manchester Woolworth's fire. This material presents a very serious hazard, because, as it burns, large amounts of highly toxic smoke, including carbon monoxide and hydrogen cyanide, are produced. It also releases large amounts of heat and will melt to form burning droplets. 'Combustion-modified' foams are now available, which will burn much more slowly with a corresponding reduction in the production of heat and smoke. The material is either used as a barrier around an ordinary foam core, or it can be used on its own.

Polypropylene stacking chairs have also produced fires which give a similar quick release of heat and smoke, due to the toxic gases produced by the polymer, and the enhanced burning characteristics of the stacking arrangement. Care should be taken to specify stacking chairs which meet ignition source 5.

Fabrics are classified by their flame retardancy. However, care must be taken in the use of artificial fabrics, which are able to call themselves 'flame retardant' as they do not burn when a flame is applied, but melt away to leave a hole which will expose the foam or filler beneath the fabric. Cotton fabrics can be treated with Proban or Pyrovatex to give good fire retardancy qualities, so that they char in the contact zone but remain in place. It is important that all fire-retardant fabrics are clearly marked with their laundry instructions, so that the retardancy is not removed by washing.

In addition to the furnishings and fittings, other of the building's contents can add to the fuel load, notably any goods stored in connection with the building's use. It is obvious that there will be increased hazards with designated warehouses or the designated storage areas within buildings, but the architect must also consider the unplanned but probable storage of goods in other spaces. The design team can play a role in controlling the positioning of fuel hazards, just as in the control of ignition hazards. The maintenance manual can also be used to limit the misuse of storage space and to ensure the separation of fuel and life risks.

The total amount of potential fuel within buildings of different types should determine the level of fire containment provided, and this will be considered in greater detail in Chapter 5 (Tables 5.1–5.3 relate likely fuel loads to the fire resistance necessary for structural elements and the maximum sizes for compartments).

2.3 Fire safety management

Larger buildings may require more than just a simple maintenance manual and architects may well find themselves involved in the preparation of the fire strategy for the building, once occupied. This strategy should already be implicit in the design, and the documentation will describe the decisions the design team have already taken on fuel limitation and ignition prevention. The fire strategy will also extend the consideration of fire prevention to include the fire safety management of the building (including communication, escape, containment and extinguishment). Buildings such as hospitals, shopping centres and large office complexes need to have such a fire strategy prepared as part of the commissioning process, and the architect should be a member of the team preparing such a document.

The fire safety strategy will set out both the normal safety procedures for the building and the action to be taken in the event of a fire. The normal safety procedures will broaden the maintenance schedules into a programme for full and regular fire safety **audits** of the building, in which all the fire safety systems and components are regularly reviewed. Such an audit allows new risks within the building to be identified and the appropriate measures taken to counter the dangers. Any large building will gradually be altered, adapted and modified. Such regular audits enable the fire safety provisions to be modified to cope with such changes. In addition, the normal fire safety procedures will specify the training which the staff require, both induction training for new staff and regular refresher training for all occupants. Such training will be more than merely 'fire drills'; it should include training in fire prevention and perhaps firefighting, as well as fire evacuation. Audits are considered in more depth in Chapter 7 (section 7.6).

The second part of the fire strategy will cover the action to be taken if ignition occurs, and the design team should be involved in the pre-planning of such action. The document will have to outline the responsibilities and duties of the staff, indicating which tactics should be attempted in what eventualities (refuge or egress, fire extinguishment or fire containment). Such a pre-planned response to a fire emergency can be used as the basis for training and should also be adjusted as the fire safety audits reveal new risks or modifications to the building in use.

Having examined the measures the architect can take to maximize the chances of preventing a fire, it must be remembered that this tactic can never be 100% effective and that ignition is bound to occur eventually. Chapters 3–6 consider the post-ignition tactics that can be attempted and which should be integrated into the building design.

Communication 3

When fire breaks out, it is essential that it is detected as fast as possible. The exponential rate of fire growth has already been stressed in Chapter 1 (section 1.2), and it is obvious that the earlier action can be taken to mitigate the consequences of ignition, the greater will be the possibility of success. Once a fire is detected, either by the occupants or automatic means, it is then necessary to communicate the location of the fire to the occupants and the fire service. The information will enable any prearranged fire evacuation to be started and trigger any automated response (such as an active smoke control system, closing fire doors or triggering a local suppression system).

It is important that designers think of the communication system as a whole rather than an isolated piece of engineering. The system should be specifically designed to relate to the nature and the form of the building, and it must provide a network from discovery of the fire to the information being delivered to each occupant, the fire service and the management.

Although a fire-alarm system is part of the communications, it is generally classified by its principal purpose as being either Life Safety (L) or Property Protection (P). Some systems may be regarded as offering both Life and Property Protection (L/P). Despite such classification, all systems will provide some alarm to initiate fire escape and to attempt fire containment and extinguishment.

3.1 Detection

Fire detection systems identify the products of a fire. For a person this is by sight, sound or smell, and for an automatic detector it is by heat, smoke, light (in the ultra-violet or infra-red wavelengths) and turbulence movement. The detection devices sensitive to these different effects will be considered separately.

3.1.1 Manual

People are probably the best 'smoke detectors' present in buildings, although there is at least one architect who claims his pet dog as the best fire detector!

People are able to recognize fire by its sound and smell and by sight, and then to quickly make a rational judgement.

The designer can make a positive contribution to fire safety in the design of circulation routes and certain types of accommodation. By achieving good passive surveillance of the building by the occupants fires will be prevented or, at least, detected earlier. This is most relevant when positioning the porter's or janitor's room, in the siting of a nursing station in an elderly people's home or hospital ward, or a supervisor's office in a large warehouse. It might be worth-while planning the circulation routes for large buildings such that all areas are kept under passive surveillance. This might mean, for example, ensuring that staff have to pass through warehouse areas on their way to the canteen. However, the desire to achieve such passive surveillance may conflict with the need to achieve fire separation (e.g. between offices and warehousing). It must also be balanced against the increased ignition risk of allowing staff into certain areas.

It is equally important that designers avoid if at all possible circulation areas which are used only for fire escape and are normally deserted. Such areas will not be under regular surveillance and a fire in such areas may not be detected quickly. They may also be a risk if they become dumping grounds for stores or junk – such materials not only posing a fire risk, but also hampering evacuation.

Fire safety training emphasizes the need to raise the alarm as the first action, but the most difficult problem for the trainer is whether to encourage people to fight a fire or leave the building. In the domestic situation a 999-call to summon the fire service and to get everyone to leave is the correct action. In more complex buildings a fire-alarm system is provided, so that an alarm can be raised by using a '**break glass**' call point (or more correctly, a **manual call point**). These call points take the form of a red box often with a clear plastic sheet etched to provide an easy break. It is usual to position manual call points on exit routes, encouraging people to leave the building and ensuring that no one has to move towards a fire in order to raise the alarm. In some buildings, such as old people's homes or hospitals, additional manual call points are required where the nursing officer or nurse/control base is located. Similarly, a night porter's position should also have a manual call point.

Good observation by staff in warehouses, factories and sports areas will contribute to fire safety through rapid detection and the early elimination of potential sources of fire. However, manual detection will only be of value while the building is occupied, and it may well be necessary in such buildings to install automatic detection to provide cover at night.

In restaurants, clubs and public houses it is wise to consider siting the call points so that they are under staff control. This design decision will need to be integrated with fire safety policy and procedures for the establishment.

3.1.2 Smoke

The most frequently used automatic detectors detect the smoke particles from a fire, and these are normally able to detect a fire at an earlier stage than heat detectors. There are two main types: ionization and optical; a dual head with both sensors is also available. The **ionization chamber detector** is the most sensitive to small smoke particles and will probably react more quickly at the early stages of a fire. The **optical detector** depends on scattering the passage of a light beam within the detector head and is therefore more sensitive to larger smoke particles.

In principle, ionization smoke detectors work by monitoring the electrical current between positive and negative plates across an air gap, in the presence of a small radioactive source. The radioactive sources cause the ionization of the molecules in the air. The ions are attracted to the respective oppositely charged plates and a modest current then flows. The introduction of smoke particles will reduce the current flow. When the current flow is sufficiently reduced or the voltage drop replicates the kind of expected performance in a fire, then the alarm is initiated. In practice, smoke detectors are usually more sophisticated than this and may compare results between a closed chamber and an open chamber. However, they are extremely sensitive to small smoke particles and are particularly effective at sensing a fire in its early stages. Ionization detector heads are usually ceiling mounted and provide coverage for about 100 m^2 of floor area.

Smoke in a light-scatter detector interrupts a beam of light from the source to a light trap and deflects some to the photoelectric cell. The current generated by the cell is monitored or has a pre-set level when it transmits an alarm signal.

Optical detectors are more adept at sensing dense smoke, while ionization detectors are more sensitive to the smaller, normally invisible, smoke particles produced at the start of a fire. Therefore optical detectors are better if smouldering fires are anticipated, and ionization detectors if the danger is flaming fires.

Care is needed in the siting of smoke detector heads and they should normally be situated at the highest point in a space. Smoke modelling may suggest other points for particular buildings with large volume spaces (e.g. churches). Locations near exhaust extracts and fresh-air inlets in kitchens and garages should be avoided. As a simple rule, it is probably better to site ionization detectors within rooms and optical detectors in corridors or on circulation routes.

3.1.3 Heat

Heat detectors are general-purpose detectors which react to a designated temperature, or rate of rise of temperature. The early heat detectors used the expansion effects in a bi-metallic strip to sense temperature, and in its simplest form a heat detector consisted of a bi-metallic strip which bends and, at a pre-determined temperature, allows an electrical contact to close which brings about an alarm signal. Many of the heat detectors currently on the market use small electronic resistors calibrated as temperature sensors rather than a bi-metallic strip. A normal operating temperature of such a heat detector head would be 68°C, and each one could protect an area of 50 m². The fixed temperature head is less likely to give false alarms where temperature variations occur normally, for instance, in laundries and kitchens. The clear disadvantage of this type of detector is in its relatively slow response time: the fire may have grown significantly before sufficient heat is generated to set off the detector.

A 'rate of rise' detector will respond more quickly to a fire, as it is triggered by a rapid increase in temperature, and is not dependent upon a particular temperature being achieved. It is particularly suitable for use where smoke detectors would prove unreliable due to dust particles, or where smoking is part of normal activity. The early rate of rise detectors incorporated a fixed temperature detector and additionally relied on the relative expansion of two different bi-metallic strips. A rapid rise in temperature caused one strip to expand to make a contact with a point held in place by a second strip shielded from rapid temperatures. The second strip was, however, allowed to expand with the ambient conditions, so a slow rise or fall in temperature would not cause a contact. The latest detectors achieve the same effect using electronic temperature sensors. A fixed temperature sensor is always included to allow for fires which have a long, slow burning development. A variety of fixed temperature settings are usually available.

The choice between the two forms of heat detector should be taken on the basis of the normal ambient working temperatures. In unheated areas a fixed temperature head will take longer to operate than a rate of rise detector, but it is to be preferred if large temperature variations recur in the normal working process. Heat detectors are particularly suitable in areas prone to smoke such as kitchens, boiler rooms, laundries and carparks.

3.1.4 Light

Infra-red radiation and ultra-violet light sensors can be used as flame detectors and they are used for more specialized detection problems. Such detectors use the radiant energy from the fire in the infra-red or the ultra-violet spectrum respectively to react on either a photoelectric cell or gas-filled sensitive tube. The infra-red detectors are usually further adapted to respond to a flickering source to avoid false alarms from electric bar heaters or sunlight. They can be used to sense through smoke, but may not pick up certain fires which burn with a clear (sometimes transparent) flame (e.g. alcohol and isopropanol). Ultra-violet light does not penetrate smoke readily, but it will be sensitive to clear flame fires. There is a danger with both infra-red and ultra-violet detectors that they may respond to very bright direct, or indirect, sunlight. Therefore the calibration and siting of detectors is most important.

Where the contents of a building are likely to produce smoke, or to smoulder before the flames appear, flame detectors are not suitable and in these situations smoke detectors are preferable. However, flame detectors are suitable in any areas where there is storage of flammable liquids in any appreciable quantity. Normally one would expect flammable liquids in use in a laboratory to be in 500 ml containers, and while several containers might be in use, accidental spillages would be limited. Larger amounts would be kept in steel bin lockers or in purpose-designed stores. In these areas flame detectors would be appropriate.

Flame detectors have also been successfully used in large volume spaces, such as churches and cathedrals, where it is difficult to predict smoke behaviour. Small fires could be picked up earlier in this type of situation as the flame flicker may be pronounced.

3.1.5 Thermal turbulence

Most of the detectors already described are point detectors, that is they detect either smoke, heat or light at a particular point or within a fixed radius of that point. Thermal turbulence detectors depend on beams which optically sense hot air currents or smoke. Interruption of a light beam indicates a fire. These detectors are extremely useful for large spaces, especially atria and bus garages. In atria they can be sited so the beams go across the large volume at varying levels. The ability of smoke to rise and then to stratify is now well known. Beam detectors can be set such that the smoke will interrupt the beams, even if it stratifies at a particular level.

Garages for public service buses can present a particular problem – smouldering cigarettes left in buses, or deliberate fire-raisers, are a familiar problem. In the garage it would not be possible to provide each vehicle with detection, but a beam detector can be used, transmitted through the windows of the buses. Really this is only possible where the vehicles form a geometric pattern when parked, and when the vehicles themselves are within a standard design range. Optical beam detectors have been recommended for use in cathedrals and large churches, where there can be problems with small fires in a large space. In this context, there may not be sufficient energy to take the smoke plume to the ceiling and a reservoir of heated air at high level gives the possibility of the stratification of the smoke. The siting of detector point or beam is a specialist task and, in such instances, assistance may be necessary from a fire safety engineer who specializes in smoke modelling.

Some detectors of this type use infra-red beams sent from a source to a receiver. The thermal effects of the fire will cause the beam to be interrupted or modulated at the receiver. Should the fire develop smoke to obscure the beam at an early stage, this will also trigger an alarm signal. The beam detector is therefore available to detect both heat and smoke. There are disadvantages to the system, the possibility of accidental interruption of the beam and of building movement. The normal maximum length of a detector beam is 100 m.

3.1.6 Smoke sampling

In some special situations, use may be made of smoke sampling systems where air is drawn in sequence from a variety of fixed locations and passed through a sensor. There may be advantages in using such a system in secure establishments where the occupants might be tempted to interfere with more conventional point detectors. The air is usually drawn along small-bore sampling pipes to a central monitoring station, which then uses smoke detectors to sense any smoke particles in the sample.

Also available is an adapted optical smoke detector known as an optical duct sensor; this can be used to sample air from the return air ducts of an air conditioning system. These detectors have to be adapted to suit the particular width of the duct. They are particularly valuable in preventing smoke spread through re-circulated air.

3.2 Comprehension/analysis

Having detected a fire, it is then a question of interpreting that signal. The various detection devices can be used to give a simple fire condition signal, while a manual call point is a simple switch which gives a signal when released. Detection devices can also be used to signal at a pre-set level of temperature or smoke level. Fire-alarm and detection systems which work in this way are termed 'conventional' systems. It was apparent to many electrical engineers that the 'conventional' fire-alarm system discarded much of the data available from the various sensors. The advent of small reliable microprocessors made it possible to analyse the results from each device in a system and obtain information on performance of detector heads to give both greater accuracy in the analysis of fire effects and an ability to predict and establish faults. These systems are termed 'addressable'.

3.2.1 'Conventional' systems

In a 'conventional' system the wiring in a particular 'fire zone' will relate to that particular fire compartment or subcompartment. The detectors will be wired such that a fire signal in that zone will give an alarm signal at the fire-alarm panel and an indication of the particular zone. In a multi-zone building a conventional alarm panel will indicate a series of zones, with the zone locations marked or a zone location diagram provided alongside the panel (Figure 3.1).

The only information provided at the fire-alarm panel is the identification of the fire zone concerned. In the event of a fire or a fault, it is necessary to go to the zone (fire compartment) to establish the source of the fire or the faulty detector. While in terms of fire containment a large compartment may be acceptable, due to a low fuel load, this might not prove acceptable in terms of locating the fire or searching the zone for survivors. In such cases, it will be necessary to

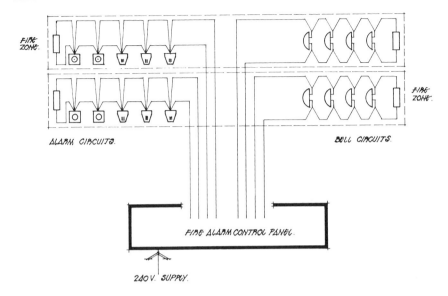

Figure 3.1 Conventional fire-alarm wiring

subdivide compartments into a series of fire zones, or use a system which provides local indicator panels with subzones.

3.2.2 'Addressable' systems

'Addressable' systems use the same number of detection devices, and alarm call points as conventional systems, but differ in that one or more microprocessors are used to control the system. These can give much more information by comparing the information received with that stored in the memory. The detector head in a 'conventional' system becomes an individual sensor on the 'addressable' system giving a precise location, and the possibility of a 'fault' reading or a fire signal (Figure 3.2).

The various components of the system can all be controlled on one wiring loop as the system in normal operation interrogates each device in turn. Micro-processor systems are inherently flexible, and can be made to scan manual call points more frequently than automatic detectors and other components. The time taken by the system to interrogate all components in the system must not be more than a couple of seconds. It is still necessary to relate the fire information to the formal 'fire zone' but the panel, in this case, will give a fire zone indication, as well as the precise component which originated the fire call. The zone definition is needed to relate the information to the fire containment measures and to any fire escape strategy.

While the capital costs of 'addressable' systems may be higher than 'conventional', the design team will need to consider their advantages in terms

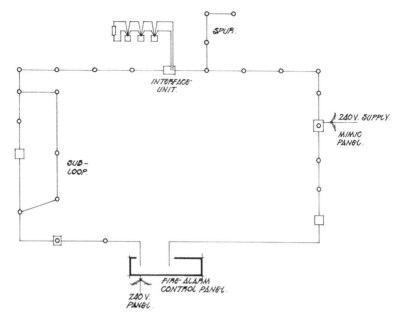

Figure 3.2 Addressable fire-alarm wiring

of installation, operation and maintenance. It is also possible to integrate these systems with a building management or security system, although it is always necessary to protect against a fire-alarm signal being affected or denied by the other system.

Some systems go one stage further and incorporate some microprocessing ability at the detector head. This is termed a 'smart' detector. These devices can be used successfully where there are regular variations in ambient conditions, by being pre-set to compare the changes with a typical change in its memory. Systems can be extended to give further signals which include faults and pre-warnings.

It is always necessary to allow for the commissioning of systems, and this is even more important with the use of 'addressable' systems. In the commissioning process, time should be allowed for the proper inspection, earthing tests, operational testing and the testing of each individual device for 'fire', 'fault' and location. It is normal practice to ask for certification of the commissioning for the benefit of the client and any inspecting authority.

3.3 Alarm

The alarm when raised is a signal for occupants to evacuate, or to be alert in preparation for evacuation. The alarm should result in the fire service being summoned, so that they can start firefighting and if necessary assist in evacuation.

It is crucial therefore that an effective system exists, and the designers should think carefully about the function of the alarm in relation to the project in question.

3.3.1 Occupants

The most common form of fire-alarm sounder used is the electric bell. It has the advantage of being universally recognized, and its sound is able to carry through a building. It can ring as an intermittent pulse or a continuous sound, perhaps signifying respectively alert and evacuation. Where a large, rambling site is involved, it may be necessary to use a siren.

Where there will be a difficulty in hearing an alarm due to disabled occupants or noisy equipment, then a visual signal, for example, flashing lights, will be necessary. The alarm may be required to wake sleeping occupants, and there have been a number of fires where lives have been lost through the lack of an alarm system. Survivors of the 1990 fire in the Cairo Sheraton hotel complained that the first indications of the fire in the hotel complex were the crackling sounds of the fire itself and the noise of fellow guests as they left. The hotel had a large number of residents, most of whom would not have been familiar with the building's layout. The fire occurred in the early hours, and the lack of an alarm meant that many were not roused until the fire was well established.

In domestic fires which develop during the night, the first warning that occupants asleep upstairs may have is the crackling sounds of a well-developed fire. Self-contained smoke detectors with integral alarms could provide an earlier warning and so significantly reduce the number of deaths.

In the variety of buildings which have resident staff, visitors, patients or inmates, the fire alarm is required to alert the staff and to commence a pre-arranged fire evacuation plan. Low-note buzzers or electronic sounders can be particularly useful in situations where it is necessary to warn staff rather than the occupants of fire.

In shops and shopping complexes the alarm may be more sophisticated than a general evacuation 'ringing'. The alarm can take the form of a coded message, for instance, asking people to leave because sprinkler testing is about to start in a particular department. This can alert staff to the existence of a fire and, more important, its location. If such arrangements are used, then fire training assumes an even greater importance.

In hospital premises and prisons it is normal to evacuate initially the fire compartment/zone of origin as part of a staged evacuation plan. Visitors are usually asked to leave the building, while the other occupants make their own way or are assisted to a safe refuge. Conventionally, in these instances, fire-alarm bells ringing continuously for evacuate, and intermittently for alert/standby, are used. However, coded messages are also sometimes used; for example, one American hospital requests 'Dr Red' to go to the department where the fire has started.

3.3.2 Fire service

For most buildings, an ordinary telephone 999-call will suffice to inform the fire service. However, the management should have a clear understanding of whose responsibility it is to make the call. In buildings with a staffed switchboard it should be done by the telephonist, and it is good practice to have the required message to the fire service ready printed for the operator to use. For the communication to be successful, the service required (Fire), the full address and telephone number need to be quoted. Even when automatic equipment is used, it is necessary to consider a 'back-up call', using the standard 999 facility.

Sometimes a direct telephone link to the fire station can be provided ('red telephone' system), but these are presently being discouraged by the fire service. There are a number of companies offering a variety of facilities to call the fire service from a central answering station.

The provision and siting of fire-alarm panels and any subpanels should be carefully considered by the designer. It must take into account the pre-planned action to be taken in the event of a fire, the need for those responsible to be informed and for the information to be readily accessible to the fire service. Where buildings are left unstaffed for periods of time, it may advisable to site panels so they can be seen from the outside.

The fire-alarm system can also be used to actuate other systems. For example, door detentes (hold-open devices), motorized ventilation dampers/controls, powered smoke extraction systems, fire extinguishing systems and staff call systems.

In the case of ventilation systems for buildings where life safety has to be considered, it is normal to allow for the automatic shut down, or change to input air with the extraction systems left running. This avoids the problem of the fire and smoke being spread by the ventilation systems. During the MGM hotel fire in Las Vegas (1980), the continued operation of the heating and ventilation system successfully spread the smoke and fire to other floors. The fire had started in the restaurant servery and it initially spread rapidly through expansion and seismic joints, and then through the air conditioning system. There was no automatic fire-alarm system, despite the building being 22 storeys high and containing a variety of high-risk areas. Although the fire was brought under control within one hour, a total of 85 people died.

The ventilation system should be so designed that the fire service can shut off the extract air or bring back on the supply and extract air from a 'fireman's switch'. This switch (or switches) needs to be in an easily accessible position. It may be necessary to separate out essential plant and non-essential plant in a large complex. Whatever the system, clear labelling is also essential.

3.4 Signs and fire notices

3.4.1 Signs

Signs are extremely important in giving good information to the occupants and the fire service; however, they should not be over-used. In instances where the communication routes and exits are obvious and the occupants familiar with their surroundings, they may be unnecessary.

Signs are essential to mark exits which are not part of the normal circulation routes. In public assembly buildings they should indicate all available routes. These signs should be rectangular with white wording on a green background; when illuminated, the contrast is frequently reversed. Any other information which relates to a safe practice or circumstance should be similarly coloured. For example, in buildings where progressive horizontal evacuation is used, doors are marked to indicate their status (Figure 3.3).

Signs which are necessary to indicate actions that should be taken are 'mandatory' signs. These signs are circular with white wording on a blue background. The most common of this type is the 'Fire door – keep shut' sign (Figure 3.4).

There may be a need to warn both occupants and firefighters of hazards within the building (e.g. radiation hazard and biohazard symbols). These signs have black wording on a yellow background (Figure 3.5).

Some signs may prohibit actions (e.g. 'No smoking'). These signs are a red circle with a central bar with a black pictorial symbol on a white background (Figure 3.6).

Red has been retained as the colour for fire equipment signs because of the obvious association of the colour red with fire (e.g. 'Dry riser' and 'Fire

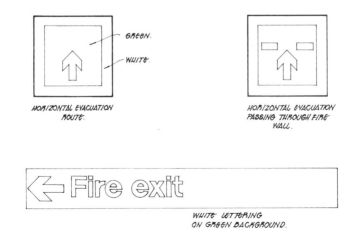

Figure 3.3 Fire escape signs: safety colour green, shape rectangular

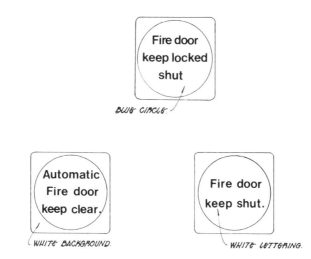

Figure 3.4 Mandatory signs: safety colour blue, shape circular

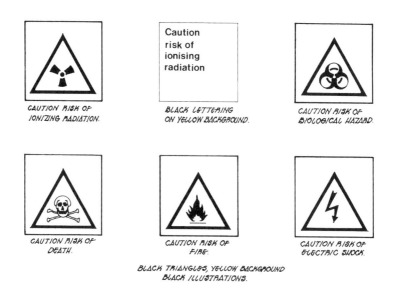

Figure 3.5 Hazard warning signs: safety colour yellow, shape triangular

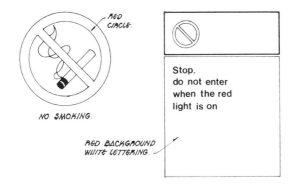

Figure 3.6 Prohibitory signs: safety red, shape circle with bar

Figure 3.7 Fire equipment signs: safety red with white border, shape rectangular

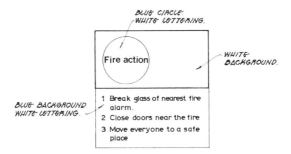

Figure 3.8 A fire safety notice

point'). The signs should be rectangular with white wording on a red background (Figure 3.7).

All signs which give vital information on fire safety should be illuminated and this should operate in both normal and emergency conditions. It is usually sufficient to ensure that light will fall onto the sign, but in situations where there are large numbers of people (e.g. theatres or sports centres) such signs will require internal illumination.

3.4.2 Fire notices

In places of work and institutions it is necessary to have a prearranged fire evacuation strategy. As well as training the staff, well-thought-out concise instructions in the shape of a fire notice need to be posted at a suitable location. If this is to be included, it should state the actions to be taken on discovering a fire and details of the evacuation plan relevant to that particular fire compartment, including any designated assembly point (Figure 3.8).

Escape

4

Every building should be designed such that the occupants can escape when fire breaks out. They must be able to reach a place of safety without being overcome by the heat or the smoke, and therefore the time needed to escape has to be shorter than the time it will take the fire to spread. This can be achieved by controlling fire spread and by ensuring that escape routes are neither too long nor too complex. The means of escape must be designed into the circulation routes within the building, and should form an integral part of the initial concept of the scheme.

It is not sufficient merely to consider means of escape as a series of protected routes whereby people can escape by their own efforts from any point in the building to a place of safety. Such a definition gives rise to a number of problems, for while many people can move adequately in evacuation, the disabled, chronic sick or sedated people will need assistance in evacuation. Therefore there are two basic escape strategies: first, **egress** – simple, direct escape from the building when the alarm is sounded; and second, **refuge** – the use of the structural fire containment of the building to provide a place of safety within the building, so that evacuation takes place from the compartment where the fire started to an adjoining compartment. Clearly, this is only acceptable when it is possible to continue further evacuation without returning through the compartment of origin.

There is also a third escape strategy which can be adopted as a last resort, namely rescue by persons from outside the building. Rescue can be considered in the case of small buildings, but it is neither reliable nor commendable. The evacuation of the occupants by ladders may have to be considered if the building has only one stairway, but this is patently not suitable for other than small numbers of people and low-rise buildings. Also because it is hard to rescue the disabled, the chronic sick and infirm, design to facilitate rescue must be regarded as a helpful feature rather than the principal escape strategy.

4.1 Occupancy

In designing the means of escape from a building the architect must consider very carefully the likely occupants and their patterns of behaviour. An understanding of the characteristics of the occupants will suggest their likely speed of travel, and in conjunction with the expected speed of fire spread, enable the architects to design adequate means of escape.

Crucial to the design of escape routes is an appreciation of the speed of fire spread (section 1.2). During the bush and scrub fires in southern France, in 1989, a French firefighter records seeing a fire overtake and consume galloping wild horses. Inside buildings fire spread can be equally startling: the film of the fire spread at Bradford City football stadium and that of the Fire Research Station's re-creation of the Stardust Disco fire both dramatically illustrate this.

The nature and numbers of the occupants is probably more influential than certain of the physical design factors emphasized in escape codes and guidance. It is the interactions of the communication system with the occupants, the effectiveness of the signposting, the clarity of the internal layout and routes, the quality of fire safety training and response that will minimize the life risk from fire. Five key characteristics of the occupants can be identified:

1. sleeping risk;
2. numbers;
3. mobility;
4. familiarity;
5. response to fire alarm.

These will now be considered.

4.1.1 Sleeping risk

Buildings where people sleep are inherently more dangerous than those used only during the daytime. This is the single most important factor to recognize for the architect involved in the fire safety design of a building. When people are asleep, there is the opportunity for a fire to grow before discovery; and even when it is discovered, the reactions of people who have been asleep will be much slower. Many a fire is simply prevented from being much more than a minor incident by prompt preventative action or by simple extinction. Moving the clothes horse farther away from the heater when things smell hot, or stamping on the small, glowing cinder that spits on to the carpet, are familiar examples from normal domestic life. The combination of the lack of preventative actions and the slow response time to a fire make sleeping risk a particular feature to be considered.

The majority of fire deaths in the UK occur in people's own homes, simply because of the added risks when asleep. The installation of a simple smoke detector would help to reduce the risk by providing some measure of warning, giving the occupants the extra couple of minutes necessary to escape.

Residential premises, whether institutional (e.g. hospitals and prisons) or commercial (e.g. hotels and guest-houses), also pose a severe risk. In institutional premises there may be staff awake and on duty all night, but detection is a valuable additional safety feature. All commercial residential premises should have detection.

The Fairfield Old People's Home at Edwalton, Nottinghamshire, was severely damaged by fire in 1974, and 18 people died. The fire was thought to have been started by one of the residents smoking in a bedroom, while most residents were asleep. It spread rapidly through the continuous ceiling void and was not detected until in an advanced stage. The greatest damage and loss of life ensued at the opposite end of the building.

The hotel fires of the late 1960s, particularly the fire at the Rose and Crown in Saffron Walden, Essex, in 1969, in which 24 died, were instrumental in ensuring the designation of hotels and boarding-houses under the Fire Precautions Act 1971.

4.1.2 Numbers

To plan an adequate means of escape the designer needs to know how many people will be in the building and where they are likely to be located. This will depend upon the building's function, but architects must remember that a building designed for one purpose may well be used for another. For example, in one Hampshire school the fire service were horrified to find that the double-height 'street' down the middle of the building was not just used for daytime circulation as they had expected, but was being used for such diverse evening activities as beer festivals.

Some buildings are designed to hold a maximum number, for example, a theatre or lecture room, but for others it may be necessary to make an assessment of the maximum numbers who might be using the space. This will be particularly important for any building involving a public assembly for work, pleasure or shopping. In the planning stage it is usual to do this by making an estimate of the likely numbers of the building type using an '**occupancy load factor**'. The area to be considered in square metres is divided by the 'occupancy load factor', giving a rough guide to the maximum numbers to be expected. The total for a large building would be calculated by adding together the totals for each individual area. No account is normally taken for the circulation areas.

Although each project should be assessed separately as part of a full safety engineering process, a simple guide of suggested 'occupancy load factors' for different building types is provided in Table 4.1. This is derived from first principles rather than any particular code or approved document and is intended primarily for student architects working at the sketch design stage. These figures provide a very rough guide and take no account of the special risks associated

Table 4.1 Building type and occupancy levels

Building type	Occupancy
1 Houses	Five times bed spaces
2 Flats and maisonnettes	Five times bed spaces
3 Residential institutions (hospitals, prisons, etc.)	Three times bed spaces
4 Hotels and boarding-houses	Twice bed spaces
5 Offices, commercial, schools	'Occupancy load factor' = 6
6 Shops	'Occupancy load factor' = 4
7 Assembly and recreation	
(a) bars	'Occupancy load factor' = 0.5
(b) dance halls, queuing areas	'Occupancy load factor' = 0.7
(c) meeting-rooms, restaurants	'Occupancy load factor' = 1
8 Industrial	'Occupancy load factor' = 5
9 Storage	'Occupancy load factor' = 15
10 Car-parks	Twice parking places

with very tall buildings (over 10 storeys) or deep basements (more than one level) which would need special attention.

In the design of flats and dwellings the normal ergonomic requirements for circulation would render most of these calculations unnecessary. However, where large numbers may occur, such estimates are essential, so that escape routes can be planned with adequate door openings, corridor widths and stairways.

In the Republic of Ireland the most serious fire in recent years was on St Valentine's Day 1981, when 48 young people died at night in the Stardust Disco in Dublin. One of the main problems was the sheer numbers of people in the disco at the time. Large numbers should not have constituted a problem, but combined with a total lack of any form of Building Regulations, inadequate local authority supervision and the untrained owners' advisers, they contributed to the disaster. The management failed to give the alarm and commence evacuation immediately the fire was noticed. Instead the occupants watched the unsuccessful attempts of two employees to extinguish the fire. Then without warning, it became an inferno, the fire spreading exceptionally rapidly due to poor seating, wall lining materials and low ceilings.

In addition to estimating the total numbers within a building, it is also necessary to identify specific areas where people will concentrate and where the density could pose problems. Large numbers of people are not going to move quickly, or even at a normal pace, so it follows that escape must be accomplished in a shorter distance. The designer may also have to consider aspects of crowd behaviour and the management of large numbers of people. As well as the adequate provision of illuminated exit routes, the clarity of direction signs and the adequacy of the communication systems, the provision of crush barriers may become important.

The potential for disaster with large numbers of people was illustrated by the crushing to death of 97 people at Hillsborough football ground in 1989. Here the surge of people trying to get into the ground trapped the helpless spectators already on the terraces against the crowd barriers and perimeter fencing, with those at the rear of the crowd having no idea of the deaths occurring at the front by the pitch. Provision for escape should have given the spectators the ability to move away from disaster, even if this had meant releasing gates in the perimeter fencing to allow access to the pitch.

Four years earlier, 55 people had died when the main stand caught fire at Bradford City football ground. Here again, the number of casualties was increased by the poor escape routes. The only rear exits were at the highest point and reached by a dimly lit, 2.6 m wide corridor, which was further narrowed by refreshment stands. Unlike Hillsborough, escape forwards onto the pitch was possible, but it was still hindered by two lines of barriers.

It is areas where a very high density of people can arise – partly through a history of tragedies with high loss of life – that are usually subject to a variety of licensing laws. Proper thought and allowance for escape at the initial sketch scheme stage can prevent abortive work.

4.1.3 Mobility

It has already been stressed that people must be able to escape from the danger areas before they are overcome by the smoke and heat from the fire. However, people will escape at different speeds and there is no ideal figure which the designer can use. Some of the occupants may be disabled, encumbered or even drunk. There have been many attempts to estimate how fast normal healthy people would move, and most come up with a figure of 50–100 m per minute. Therefore the figure of 50 m per minute should, perhaps, be used as a very crude and cautious guide to likely speed of movement for the able-bodied.

The worst cases will be those where assistance is essential if they are to move at all. Further complications can occur when considering orthopaedic patients with full traction equipment, or intensive-care patients on life-support systems where it is necessary to carry out the evacuation and sustain the support systems. In between these two extremes lies the majority of the population with limited mobility. The designer will need to assess what proportion of occupants may be unable to move away readily from a fire and plan accordingly.

The multiplicity of handicap is enormous and can be considered as a spectrum which includes the mentally handicapped, the partially sighted and the deaf or hard of hearing, as well as those with physical handicaps. In this context, it may be necessary in the design of enclosed shopping complexes to provide way-finding maps for the partially sighted and to arrange for fire wardens to assist in evacuation according to a predetermined fire strategy.

It may be pertinent in the design of very large schemes or large spaces to seek the assistance of computer modelling techniques. In the development of the Stansted Airport Terminal building, Ove Arup and Partners were commissioned to develop a fire engineered solution. This included a portrayal of smoke spread from a 'standardized' fire using a computer model and imposing a random distribution of individuals with different movement capabilities. The modelling is able to demonstrate the feasibility of the last disabled person struggling through the exit before the black cloud engulfs the area. The modelling is a tool to enable the designer to establish the suitability of large volume areas and escape, therefore, one aspect of a first principles approach.

Speed of escape is also effected by the design of the escape routes. Different parts of the route (through a room with furniture, along an unobstructed corridor or down stairs) will mean people move at different speeds. In most buildings it is necessary to size the circulation and escape routes to accommodate the widest expected traffic. Due allowance needs to be made for the handicapped, though it should not be assumed that all handicapped persons are wheelchair-bound.

Other types of problem might occur with mobility where the need for security is high. In prisons or in hospital secure units it may be necessary to think in terms of the occupants being assisted to escape or being able to escape only into secure compounds. This can, in turn, lead to additional problems for the firefighters. In one incident a fireman attacked a window in a secure unit with a small axe to gain entry, only to have the axe bounce back into his chest. A proper fire routine, escape and firefighting strategy is essential for such establishments. These matters need to form part of the design brief and sketch design of any project.

4.1.4 Familiarity

If the occupants are familiar with a building, they will obviously find less difficulty in escaping from fire than those unfamiliar with their surroundings. In a strange building people will instinctively try to escape the way they came in, and it may be hard to persuade them to escape via 'official', designated escape routes if these are in the opposite direction. Therefore normal circulation and exit routes should always be regarded as escape routes. Escape routes which are not normally used, and only available in an emergency, should be avoided if at all possible. If they are unavoidable, then they will need explicit signposting.

Familiarity will vary with building type. In a normal domestic situation the occupants will be very familiar with the layout of their own house or flat. Similarly, offices and factory buildings will probably have a stable workforce that will be familiar with access and exit routes. Problems are likely to occur, however, in places like hotels and hostels where the residents may only stay a short time. The problem is particularly severe in clubs and cinemas where the occupants can be unaware of which level they are on, as well as the location of exit routes.

In the fire at the Summerland leisure complex in 1973 the high death toll (50 out of 3000 in the building) was, in part, due to the complexity of the building and the unfamiliarity of the occupants. It had seven levels, built against a cliff face and the main entrance was on level four. The building had been conceived as the solution to the British summer, creating a Cornish village with a Mediterranean climate. By the time it opened in 1972, it had become more of a fun palace, with bingo, disc-jockeys, amusement arcades and a variety of licensed bars. The Committee of Inquiry concluded that the high loss of life was due to rapid fire spread, delays in evacuation due to management faults, lack of escape stairs, the locking of exit doors and the confusion as parents and children in different parts of the building tried to find one another. Evacuation was hampered by the wrong response and by the lack of familiarity with the building's layout, some vertical escape stairs not being used by the occupants because they did not realize they were there.

4.1.5 Response

The likely response to a fire or a fire alarm has to be considered as another feature. When a fire occurs or an alarm sounds, a variety of actions may take place. In a building where there is a well-disciplined staff with a planned evacuation strategy, the response will be markedly different from buildings which contain people who may be unwilling or unable to appreciate the danger. In this context, an office building or an acute/surgical hospital may be safer examples, while a house or flat which is completely familiar could conceal hazards. In high ignition or fuel hazard premises there may also be the problem of isolated staff who could be unaware of a developing fire.

There have been various studies of human response to a fire or fire alarm. These show that individuals often do not react immediately to a fire alarm; instead they may make contact with others in the search for further information, they may make the wrong assessment of the situation or they may just ignore the alarm. Studies of past fires show that panic is not frequent, but people often make mistakes or do not have sufficient information to respond effectively. Where there is a clear evacuation strategy, or where there are clear instructions or signals, people will follow them.

The inclusion of a fire-alarm and detection system can be critical in ensuring the correct response and encouraging swift action. Training is also crucial in eliciting the right response from the staff or occupants. In the Taunton sleeper train fire (1978) there were survivors from the sleeping compartments close to the source of the fire, while others who were farther away died. Subsequent investigation showed that one of the survivors had some training in fire safety and used that knowledge in effecting her escape by crawling along the corridor. On finding the carriage door locked, she returned to her compartment, shut the door and waited until she heard the fire brigade enter before attempting again.

4.1.6 Occupancy and building type

Table 4.2 offers designers, particularly student architects, general guidance on making an assessment of the critical factors discussed in the text related to building type. Some building types or some multi-use complexes mean that only a functional, holistic approach to fire safety will be appropriate. The table can

Table 4.2 Building type and occupancy characteristics

Building type	S	N	M	F	R
1 Houses	x	–	–	–	x
2 Flats and maisonnettes	x	–	–	–	x
3 Residential institutions (hospitals, prisons, etc.)	x	x	x	–	–
4 Hotels and boarding-houses	x	–	–	x	–
5 Offices, commercial, schools	–	x	–	–	–
6 Shops	–	x	–	x	–
7 Assembly and recreation (theatres, cinemas, etc.)	–	x	–	x	–
8 Industrial					
(a) high ignition hazard (oils, furniture, plastics)	–	–	–	–	x
(b) medium ignition hazard (garages, printing, textiles)	–	–	–	–	–
(c) low ignition hazard (metal working, electrical, cement)	–	–	–	–	–
9 Storage					
(a) high fuel hazard	–	–	–	–	x
(b) medium fuel hazard	–	–	–	–	–
(c) low fuel hazard	–	–	–	–	–
10 Car-parks	–	–	–	–	–

x = a problem in escape.
S = sleeping.
N = numbers.
M = mobility.
F = familiarity.
R = response.

therefore be used to establish a risk factor for any type of occupancy, or any mix of occupancy. It must be emphasized that this approach may not be endorsed by the various enforcing authorities in various countries throughout the world, but it will offer the designer some kind of rationale on which to base a sensible proposal.

The table shows that the building type posing the most severe risks is residential institutions, which show problems under three of the five factors. Domestic properties (houses, flats and maisonnettes), hotels and boarding-houses, assembly and recreation buildings, and shops have problems under two headings. All other building types do not pose such a severe risk, and except for high ignition and high fuel hazard premises, they do not register any severe problems on the table. The special risks associated with very tall buildings (over 10 storeys) or deep basements (more than one level) would, of course, need special attention.

4.2 Travel distances

Having considered the characteristics of the occupants which will influence escape, and having seen how these can be related to building type, it is important to look at the stages in the process of escape and the maximum distances people can be expected to travel. Escape is generally considered in four distinct stages.

Stage 1 – escape from the room or area of fire origin.
Stage 2 – escape from the compartment of origin by the circulation route to a final exit, entry to a protected stair or to an adjoining compartment offering refuge.
Stage 3 – escape from the floor of origin to the ground level.
Stage 4 – final escape at ground level.

In a very simple layout the stages are as illustrated in Figure 4.1. However, in more complex buildings where the evacuation is phased, stage 2 of the evacuation may only be to evacuate to a place of safety on the same level (a refuge).

In high-rise buildings it may not be practicable or desirable to commence total evacuation and instead the design must allow for a phased evacuation. Reliance

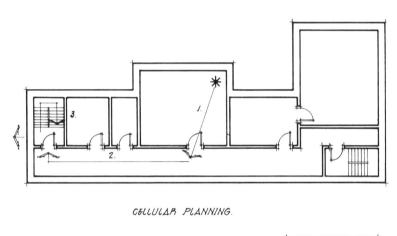

CELLULAR PLANNING.

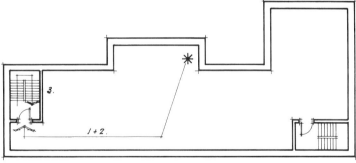

OPEN PLANNING

Figure 4.1 Stages of escape

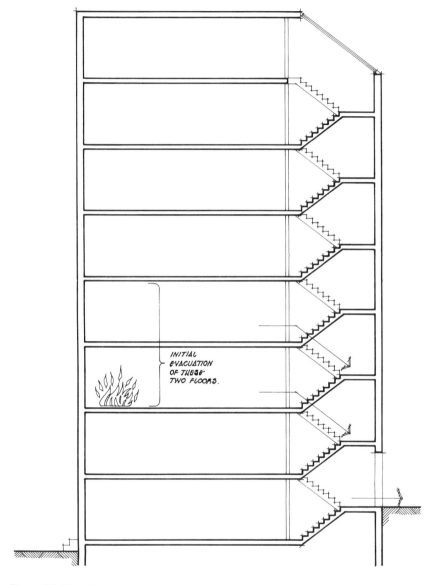

Figure 4.2 Phased escape

is placed on structural fire precautions to restricting fire spread, so that those most at risk can evacuate first, while the others can be evacuated later, once the fire-fighting teams arrive. This type of scenario clearly demonstrates the interaction of escape, containment and extinguishment, with communication in the central role. **Phased evacuations** may involve stages 1 and 2, with stage 3 held in reserve, the total evacuation of some occupants (stages 1–4), while others are put on 'alert' in readiness for evacuation (Figure 4.2).

4.2.1 Stage 1 (out of room of origin)

In achieving escape from a room the speed of the fire spread needs to be considered and compared with the speed with which the occupants can leave. However, it is extremely difficult to predict the rate of fire growth or fire spread – all that can be done is to try to ensure that the occupants of the room become aware of the fire as early as possible. In a large room it may also be necessary to provide more than one exit, so that the occupants are never too far from the door. The maximum distance they should have to travel is referred to as the stage 1 travel distance.

It is sometimes necessary to plan an area such that access to one room is through another (Figure 4.3). It may be a requirement of the client that a general office provides access to inner offices or that secure areas, and clean laboratories have to be entered through outer rooms. In fire safety terms, this layout is referred to as an 'inner room' and an 'access room'. While the whole area is treated as part of stage 1 escape, it is necessary to ensure that the occupants of the inner room are aware of conditions in the access room. Commonly this is done by designing sufficient glazing in the wall and door separating the two. In photographic developing rooms this may not be possible, and other systems are then needed to alert the occupants of the inner room at an early enough stage not to prejudice the escape.

Similarly, it may be necessary to restrict the use of the access room to prevent the escape route from the inner room being an unacceptable risk by way of its contents or its likelihood of having a fire.

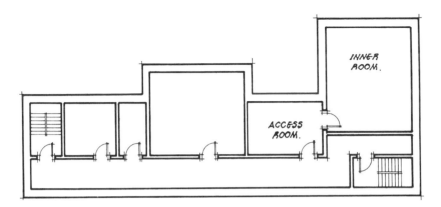

Figure 4.3 Inner room and access room

4.2.2 Stage 2 (out of compartment of origin)

The next stage of the escape process is to leave the compartment or sub-compartment where the fire ignited. This is usually by the circulation route which can lead to the outside or to a protected stairway or to an adjoining compartment which provides a refuge. The occupants need to reach a place of safety away from the area affected by fire, either by leaving the building entirely or by moving away from the fire protected by sufficient fire containment measures. In this context, the fire containment may include subdivisions of fire com-partmentation termed '**subcompartments**'. Compartments are defined by their ability to give one hour's protection from fire, while subcompartments only provide half an hour (section 5.2).

The designer should design the compartmentation of the building to ensure that all the occupants will have sufficient time to be able to escape from the compartment of origin before being overcome by fire and smoke. However, as people, buildings and fires vary so much, it is extremely difficult to calculate precise times. Therefore most codes and legislation specify distances derived from past fires and past 'experience'. The most commonly quoted of these figures is two and a half minutes, and this has often been taken as a design guide. However, its authority rests more upon legend than fact. During a fire at the Empire Palace theatre in Edinburgh, in 1911, the whole audience is reported to have been able to evacuate in the time it took the orchestra to play 'God Save the King' (2½ min), even though ten people died backstage.

Even if it was possible to establish reliable escape times, these would still have to be translated into escape distances, even more approximations arising. The architect is therefore faced with only two options, either to embark on a full mathematical analysis of all likely fire sites and types of fire and to simulate their spread and effect on the likely occupancy, or to work from first principles and make a reasonable assessment of the life risk and feasible travel distances. Obviously, it will be necessary to comply with the relevant local and national legislation, but at the early design stages it is more important to get the principles at least approximately correct (Figure 4.4).

Table 4.2 linked occupant characteristics to building type, and this first principles approach can be developed to give rough guides to reasonable travel distance. These figures do not follow any particular code or approved document and are intended primarily for student architects working at the sketch design stage.

The combined distances in Table 4.3 give a suggested overall distance. All occupants, wherever they are situated in the building, should not have to travel more than this distance to reach a place of safety. These figures have been developed by taking a maximum travel distance of 50 m for stages 1 and 2, and then reducing this by 10 m for each problem identified. Therefore a building type with two of the five characteristics identified as problems would be reduced to a combined travel distance of 30 m for stages 1 and 2. The building types with the most problems (three out of five) are residential institutions, and these have

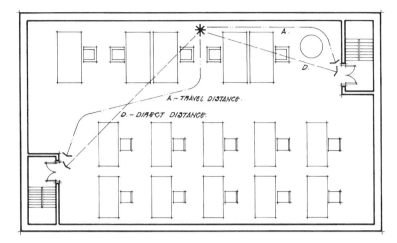

Figure 4.4 Direct distance and travel distance

Table 4.3 Building type and travel distances

Building type	Travel distance (m)		
	stages 1 + 2	stage 1	stage 2
1 Houses	30	15	15
2 Flats and maisonnettes	30	15	15
3 Residential institutions (hospitals, prisons, etc.)	20	10	10
4 Hotels and boarding-houses	30	15	15
5 Offices, commercial, schools	40	20	20
6 Shops	30	15	15
7 Assembly and recreation (theatres, cinemas, etc.)	30	15	15
8 Industrial			
(a) high ignition hazard (oils, furniture, plastics)	40	20	20
(b) medium ignition hazard (garages, printing, textiles)	50	25	25
(c) low ignition hazard (metal working, electrical, cement)	50	25	25
9 Storage			
(a) high fuel hazard	40	20	25
(b) medium fuel hazard	50	25	25
(c) low fuel hazard	50	25	25
10 Car-parks	50	25	25

a combined travel distance reduced to 20 m. Offices with only one problem have a travel distance of 40 m.

The initial figure of 50 m has been chosen as a conservative estimate of what an able-bodied adult can walk along an unobstructed corridor in 1 min. By selecting the distance capable of being travelled in 1 min it is being stressed that, ideally, stages 1 and 2 should be completed within 1 min of the occupants becoming aware of the fire, either by detection of it themselves or hearing the alarm. This figure of 1 min is as open to criticism as are all other estimates, and it is only offered as the basis of a very rough guide to travel distances.

The combined distances can be separated to provide travel distances for stage 1 (out of the room of origin) and stage 2 (out of the compartment of origin). The distance of stage 1 is taken as being half of the total distance. Stage 1 escape, from the room of origin, will almost always be possible only in one direction and so the figure given for stage 1 escape is based on a single route. However, once outside the room of origin, the designer must always provide alternative routes, so that the occupants can turn their backs on the fire to make their escape. Therefore the stage 2 distances are based on the assumption that there are alternative escape routes and the distance given is that to the nearer of the exits (Figure 4.5).

The figures in Table 4.3 provide a very rough guide and take no account of the concept of equivalency, outlined in Chapter 1. It might well be that if alternative fire safety measures are included in the design (e.g. auto-suppression), then the travel distance may perhaps be able to be increased. The special risks associated with very tall buildings (over 10 storeys) or deep basements (more than one level) would need special attention.

For projects which defy easy classification it is suggested that human behaviour characteristics could give a risk factor which, by application, would give an appropriate guide to escape distances suitable as a basis for sketch design.

Tables 4.4 and 4.5 have been included to provide very rough guidance, again primarily for the student architect working at sketch design stage. They provide information on the number of exits which will be necessary from large spaces (e.g. concert halls, exhibition spaces, etc.), and on the minimum exit widths for doors and stairs for different numbers of people. They must be used in

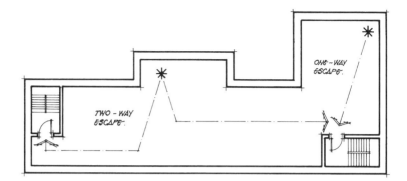

Figure 4.5 One-way and two-way escape

Table 4.4 Minimum numbers of exits from large spaces

No. of people	No. of exits
1–50	1
51–500	2
501–1000	3
1001–2000	4
2001–4000	5
4001–8000	6
8001–12 000	7

Table 4.5 Minimum total widths of escape routes and exits

No. of people	Total width of exits (mm)
1–50	800
51–100	900
101–180	1000
181–200	1100
201–220	1200
221–240	1300
241–260	1400
261–280	1500
281–300	1600
301–320	1700

conjunction with the travel distance figures provided in Table 4.3, and with the same caution.

4.2.3 Concept of 'refuge'

Stage 2 of escape ends when the occupants reach an adjoining compartment for refuge, or a protected stairway which will lead them to the ground level. Descent to ground level is stage 3 escape, but this is not always necessary or desirable. It may be more sensible to take refuge in another part of the building separated from the fire by a series of compartment walls (Figure 4.6). The codes of practice for escape for disabled persons and guidance for homes for the elderly, and the guidelines for new hospitals and existing hospitals, all recognize the importance of the concept of refuge. At its simplest, it means that the disabled infirm or those of limited mobility move only a short distance within the building to a fire compartment or protected zone to await further evacuation if the fire is not brought under control. In the case of the disabled in shops and offices, the concept is limited to the provision of fire-protected 'refuge' areas on each floor level adjoining the lifts and stairs which serve the building. In other premises the buildings are subdivided into protected areas of subcompartments which act both for fire containment and as reception areas for those evacuating (Figure 4.7). Movement to refuge is sometimes called **progressive horizontal evacuation**.

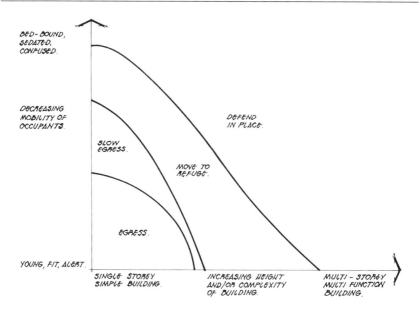

Figure 4.6 Egress versus refuge

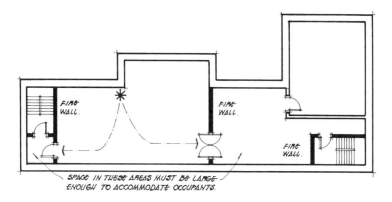

Figure 4.7 Horizontal escape to refuge

4.2.4 Stage 3 (out of the floor of origin)

Having escaped from the compartment where the fire originated, the occupants will need to reach ground level, if they are not already there, and this vertical escape is classified as stage 3. Even if the evacuation plan involves the tactics of refuge on the floor of origin, vertical evacuation might still be necessary as a last resort.

The significant feature of stage 3 escape is that those escaping should have achieved a position whereby they are protected from fire. They should then be able to leave the building not by going through any further fire hazard, but by a route to a place of safety outside the building. In larger buildings the evacuation is 'phased', so that all the occupants do not arrive at the staircase at the same time.

For most of the occupants vertical evacuation will be by staircases and normal lifts would never be used. Using lifts in a fire situation is dangerous, as the occupants may be trapped or taken to a floor which is at risk. Even waiting for a lift is dangerous, as it delays escape. However, for the severely disabled it may only be possible to evacuate using a specially designed lift. This will need an independent electrical supply, the potential for fire warden control, and communications between the lift car and the evacuation control point.

4.2.5 Stage 4 (final escape at ground level)

Stage 4 escape is from the foot of the staircase to the outside. The stairs should not all converge into one common area at ground level, otherwise a single incident can simultaneously block all routes. Although this might appear obvious when stated here, there have been a number of examples of buildings where protected routes from upper floors discharge into a common central atrium.

It should not be forgotten that the **final exit** and external design of a building also have to be considered in escape planning: it must be possible to leave the building and reach a **place of safety**. Planning needs to take account of the volume of people that may escape from a building, and their need for a readily identifiable assembly point or transfer area. Where large numbers of people may be involved, it will be necessary to plan these areas, so there is no conflict of use as the emergency services arrive and begin to fight the fire.

4.3 Rescue

Buildings divided into several fire compartments at each level, and with escape stairs positioned such that from no area is there only one direction of escape, should not have to rely on rescue coming from outsiders. However, many buildings (including most domestic dwellings) are single-staircase buildings, while others either do not have any fire compartmentation or have inadequate standards of fire separation. In these situations, the occupants may need to be rescued from the face of the building or from balconies. In designing to provide refuges for the disabled or infirm there is also the assumption that rescue will eventually be provided by some external group, whether by the fire service or staff from elsewhere in the building.

There are two aspects to designing to make rescue easy. First, it must be possible for the fire service to get close to the building. With low-rise housing, it might be sufficient for a pumping appliance to get within about 40 m of the front door, with access for ladders. With larger buildings, any rescue will be more limited and may only be possible by hydraulic platform – for this to be feasible there has to be good vehicular access to the face of the building (Figure 6.2).

Second, the window openings must be designed to permit access by the fire service from the outside. It is possible to design windows which allow normal, fit people to leave the building, but sometimes this can conflict with the security needs of the building. In the Woolworth's fire in Manchester, in 1979, people were trapped behind barred windows from which escape was impossible. Window escape must always be regarded as an opportunity for rescue rather than a planned escape route, and balconies will provide a much safer place to await rescue. In some situations, balconies can also provide the fire service with a route of entry into a flat without using the front door which might compromise escape from the flats above.

4.4 Escape lighting

It may be necessary to light escape routes such that they can be used in a fire when a failure of a local electrical circuit is probable. In general, this will be required in all buildings with the exception of low-rise housing. It is of paramount importance in assembly buildings, for instance, in theatres, cinema, clubs and discos. In one notorious incident the Six-Nine Discothèque in La Louvière, Belgium, 15 of the 60 people present lost their lives on New Year's Eve in 1975. There were inadequate means of escape, no emergency lighting and the club was decorated for the season with flammable decorations. When the partygoers lit matches and lighters in order to see, they accidentally ignited the decorations. This led to a rapid spread of fire and the loss of life.

Escape lighting must be distinguished from the **emergency lighting** which might be provided, on failure of a mains supply, by a standby generator. Such emergency lighting probably will not function in a fire due to local circuit failure and escape lighting must be provided by self-contained fittings which are capable of running for a set period of time, usually 3 h. These may be separate fittings or incorporated in the normal lighting units. If it is possible to install the escape lighting at low level, then this is even more useful as corridors and rooms will fill with smoke from the ceiling downwards, and lights below this smoke level are more necessary.

The designer should provide escape lighting for the circulation routes (stage 2), stairways (stage 3) and final exits (stage 4). Stage 1 escape routes, from the room of origin, will only require illumination where there are more than 50 occupants or there is a particularly tortuous route, for example, in staff changing-rooms. It is particularly important to light changes of direction in circulation routes and to indicate the location of relevant equipment, for example, hosereels or switchgear.

In industrial buildings and in plant areas it may be appropriate to use a photoluminescent paint to denote escape routes and final exits. It could also be of use in large semi-outdoor stadia as a form of way-marking that can be seen in dim conditions or at night.

Containment

5

The ability of a building's design to contain a fire once started is critical to the protection of the property, the lives of the occupants, and also to the surrounding people and buildings. It is the fire mitigation tactic most clearly covered by legislation and also the one with which the insurance companies are most concerned.

Whether or not a fire is detected and the communications system alerts people and equipment to take countermeasures, the design of the building should be such that the fire is contained and limited. Fire containment should be the 'fail safe' tactic which the designer has provided, even if all other measures are ineffectual. As such, it is the one most attractive to regulatory and legislative authorities, and the one they are least happy to see involved in 'trade-offs' or calculations of equivalency.

Fire containment provides the opportunity of achieving both of the fire safety objectives: property protection and life safety. That is, property protection through the limitation of fire spread and fire resistance provided to the elements of structure, and life safety through the limitation of smoke spread and through the provision of places of refuge within the building to which the occupants can retreat. This concept of refuge, as we have already described in Chapter 4, is particularly important in situations where evacuation or escape from the building is going to be hazardous or very time-consuming. Hospital design is now based around the concept of progressive horizontal evacuation to places of refuge, evacuating seriously ill patients to the outside might cause more life loss than the fire they are seeking to avoid. In high-rise buildings the fire must be contained and eventually extinguished from within the building because the fire may be below the occupants and the distances to ground level might be too great.

As always, it is the heat which is most dangerous to the building structure and the smoke which is most dangerous to the occupants. It is necessary that containment measures tackle both these risks and stop the spread of both smoke and heat (Figure 5.1).

Fire containment is not only about containing the fire products (flame and smoke) to a particular part of the building of origin. It may well be essential

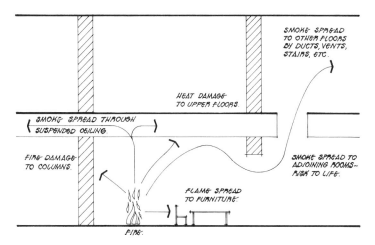

Figure 5.1 Threats from heat and smoke

also to prevent the fire spreading to adjoining properties and starting a more general and serious conflagration. In this instance, it is the flame which is most dangerous, and risks of fire transmission through radiant heat, and through the spreading of burning particles from the building carried by convection currents, that must be designed against.

It is possible to design both **active** *and* **passive fire containment** measures. Active measures are those which require some form of communication to occur by informing people or equipment of the presence of the fire and instructing them to take measures to contain its spread. Most active measures of containment are concerned with the control of smoke spread and rely on the detection of the fire triggering some form of countermeasure. These will be considered in the fourth and final sections of this chapter. The most common active fire safety measures are probably sprinklers and other forms of auto-suppression; these will be considered in Chapter 6 as they are concerned with fire extinguishment as well as containment.

Passive measures of fire containment concern the nature of the building struc-ture, subdivision and envelope. They will last the life of the building and will always be available as a defence against fire spread. Such passive measures can be considered under three headings, as follows.

1. **Structural protection** – the protection against the effects of heat provided to the structural elements of the building: columns, loadbearing walls and floors.
2. **Compartmentation** – the division of the building into different areas and the resistance to fire and smoke offered by such subdivision: internal walls, doors and floors.
3. **Envelope protection** – the protection offered by the envelope of the building to both the surrounding properties and people (from a fire within the building)

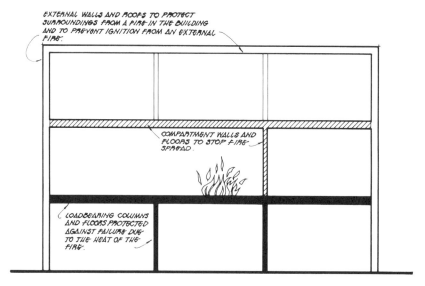

Figure 5.2 Passive fire containment

and the building itself and its occupants (from a fire in adjoining property): external walls and roofs.

These will now be considered in the first three sections of this chapter under the relevant headings.

5.1 Passive measures: structural protection

5.1.1 Protection of structural elements

The level of fire protection which it is appropriate to give to the **structural elements** will depend upon the need for escape and extinguishment. First, how long will escape from the building take, and does the safety of the occupants depend upon the provision of places of refuge inside the building? Second, is it necessary for firefighters to work inside the building, and is it necessary that the structure survives, so that the building can be rebuilt after a fire?

If the building only has to survive until all the occupants have been evacuated, then the necessary structural protection might have only to last a short time, perhaps half an hour. However, if the life safety strategy relies on the provision of places of refuge within the building, or it is necessary for firefighters to be able to work safely in the building, then protection requirements might rise, certainly to 1 hour and perhaps more. It might also be important to the building's insurers that repair rather than rebuilding is possible, and this might lead to fire protection times of 2 hours or even 4 hours.

Table 5.1 Building type and fuel load

Building type	Fuel load
1 Houses	Low
2 Flats and maisonnettes	Medium
3 Residential institutions (hospitals, prisons, etc.)	High
4 Hotels and boarding-houses	Medium
5 Offices, commercial, schools	Medium
6 Shops	Medium
7 Assembly and recreation (theatres, cinemas, etc.)	High
8 Industrial	
(a) high fuel load (oils, furniture, plastics)	Very high
(b) medium fuel load (garages, printing, textiles)	High
(c) low fuel load (metal working, electrical, cement)	Medium
9 Storage	
(a) high fuel load	Very high
(b) medium fuel load	High
(c) low fuel load	Medium
10 Car-parks	Low

The amount of fire resistance which must be provided will depend upon the fuel load of the building. To provide a very rough guide building types can be grouped accordingly, as in Table 5.1.

Although each project should be assessed separately as part of a full safety engineering process, Table 5.2 provides a crude guide of suggested periods of fire resistance (in minutes). This is derived from the first principles approach in Table 5.1, rather than any particular code or approved document. Very high fuel loads are considered to need up to 120 min of fire resistance, depending on height. High fuel loads are considered to need between 30 and 90 min depending on height. Medium fuel loads are considered to require only 30 min of fire resistance, except above two storeys where 60 min is more appropriate. Buildings with a low fire load only need 30 min fire resistance above ground floor, and nothing if single storey.

Table 5.2 is intended primarily for student architects working at the sketch design stage. The figures offer a very rough guide and take no account of the concept of equivalency, outlined in Chapter 1. The special risks associated with very tall buildings (over 10 storeys) or deep basements (more than two levels) would need special attention.

Having established the length of time for which the structure of the building must survive the effects of heat, it is then possible to design the structural elements to provide this amount of safety. However, structural protection is only as good as the weakest point in the design, and it is essential that detailing of the junctions between structural elements is as good as the fire resistance of the elements themselves.

An additional problem for the structural elements in a fire is that the progressive collapse of the building can increase the loading which they have to carry.

Table 5.2 Building type and fire resistance

Building type	Fire resistance (min)		
	1	*2*	*3 or more storeys*
1 Houses	0	30	30
2 Flats and maisonnettes	30	30	60
3 Residential institutions (hospitals, prisons, etc.)	30	60	90
4 Hotels and boarding-houses	30	30	60
5 Offices, commercial, schools	30	30	60
6 Shops	30	30	60
7 Assembly and recreation (theatres, cinemas, etc.)	30	60	90
8 Industrial			
(a) high fuel load (oils, furniture, plastics)	60	90	120
(b) medium fuel load (garages, printing, textiles)	30	60	90
(c) low fuel load (metal working, electrical, cement)	30	30	60
9 Storage			
(a) high fuel load	60	90	120
(b) medium fuel load	30	60	90
(c) low fuel load	30	30	60
10 Car-parks	0	30	30

If the basement of a building is intended to survive, then in designing the required level of fire protection it is necessary to consider the additional loads that might be imposed upon the basement with the collapse of floors above (Figure 5.3).

It is also important that in complex structures all critical components are given an equivalent level of fire protection. For example, in a single-storey,

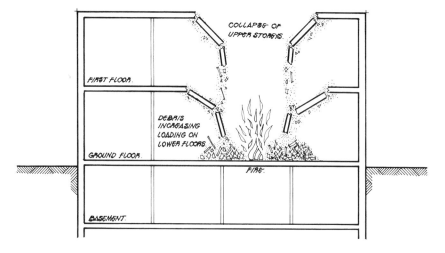

Figure 5.3 Collapse of upper storeys

steel-framed building the failure of the roof may remove the necessary lateral bracing and the result may be consequent collapse of the main structure, even though that structure is well protected against fire.

5.1.2   Fire resistance

The ability of a structural element to continue to function when subjected to the effects of heat is defined as its fire resistance and this is normally measured in terms of time. It is the fire resistance of the assemblies, not just components, which must be evaluated.

We have already discussed some of the problems of terminology and fire tests in Chapter 2, and fire resistance has been described as one of the essential terms with which architects must become familiar. The fire resistance of a component, or assembly of components, is measured by the ability to resist fire by retaining its loadbearing capacity, integrity and insulating properties (Figure 5.4). The **loadbearing capacity** of the assembly is its dimensional stability. The **integrity** of the assembly is its ability to resist thermal shock and cracking, and to retain its adhesion and cohesion. The **insulation** offered by a material is related to its level of thermal conductivity. Fire resistance is normally defined under these three characteristics (loadbearing capacity, integrity and insulation) and given in minutes or hours of resistance.

In the case of elements of structure, then only stability and integrity are immediately essential; however, if the element of structure is also acting to sub-divide the building either horizontally (floors) or vertically (walls) to contain the fire, then the insulation is also important.

When considering the fire resistance of a structural assembly, the designer

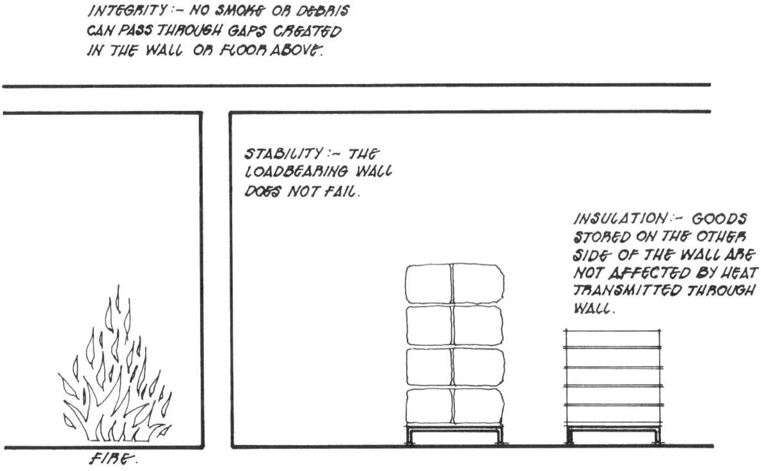

Figure 5.4 Stability, integrity and insulation

must be aware that there can be significant differences between the performance of assemblies under test conditions and in reality. Obviously, test samples are of the highest quality and workmanship; these standards will have to be repeated on site if the same level of fire resistance is to be achieved. Test samples are also new, and it is important that the fire resistance of the assemblies in place is not jeopardized by the effects of mechanical damage, weathering or thermal movement. The designer must be aware of what is likely to happen over the life of the building and make allowances for this at the design stage.

The next five sections will consider the fire resistance of the materials most commonly used by the architect. Some of these have an inherent fire resistance, with others the designer must take steps to improve their fire resistance in certain conditions; there are three principal methods of doing this.

1. **Oversizing** – deliberately increasing the size of an assembly, so that part of it can be destroyed without affecting the structural performance of the rest.
2. **Insulation** – providing a layer of insulating materials around the assembly to protect it from the heat of the fire.
3. **Dissipation** – ensuring that heat applied to the assembly is rapidly dissipated to other materials or to the air, so that the temperature of the assembly is not raised to a critical level.

5.1.3 Wood

Wood burns, but because it burns at a regular, measurable rate it is possible to deliberately oversize timbers, so that they can be used as structural elements. Such oversizing is often described as '**sacrificial timber**'. The surface degradation of the wood is normally in the form of charring, and flaming will occur only with temperatures at the surface in excess of 350 °C and the presence of a pilot ignition source. As the outer surfaces of a timber member char, they tend to stay in place and the inner core of wood remains relatively unaffected and can retain its stability and integrity (Figure 5.5). The rate of charring may vary from 0.5 mm min^{-1} (oak, teak) to 0.83 mm min^{-1} (western red cedar), but a value of 0.67 mm min^{-1} is a widely accepted estimate for structural species. This approximation applies both to solid members and laminates, though laminates may actually perform better as they will not be so prone to knots or other deformations of the timber. The use of flame-retardant treatments will not normally slow down the charring rate. It is, of course, possible to protect timber by the use of insulating materials, but as the choice of timber has probably been made because the designer wishes it to be exposed, this is an unattractive option in new buildings. However, it may be necessary when improving the fire safety of existing timber structures to consider the cladding of timber elements with insulating materials. The great advantage of timber to the designer is that failure is predictable and will occur slowly, the great disadvantage is the dramatic increase in the cost of timber elements which have had to be deliberately oversized.

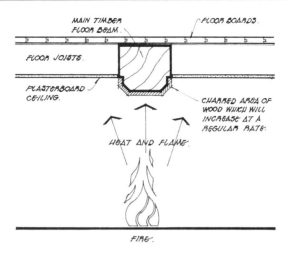

MAIN TIMBER
FLOOR BEAM.

FLOOR BOARDS.

FLOOR JOISTS.

PLASTERBOARD
CEILING.

CHARRED AREA OF
WOOD WHICH WILL
INCREASE AT A
REGULAR RATE.

HEAT AND FLAME.

FIRE.

Figure 5.5 Charring of wood

5.1.4 Steel

Unprotected steelwork will lose approximately half of its strength in temperatures of 500–550 °C and is therefore very vulnerable in a fire. As a result it is essential that steel structural assemblies are protected either by insulating materials or by the dissipation of the heat on the steel.

There are a variety of insulating materials for steelwork. Other structural materials (e.g. brick or concrete) can be used, but this is a very expensive solution. The more common materials are insulating boards, sprayed coatings or intumescent paints. Insulating boards can be used to encase steel beams and columns; there are also available insulating sheet materials which can be used to protect whole walls. The technology of their use is well documented, but care must be taken in the detailing of all junctions to ensure that no areas of unprotected steelwork are exposed. The disadvantage for the designer is the added bulk which this encasement of steel means, and the care which must be taken to ensure proper construction. Sprayed coatings are normally of mineral fibre or vermiculite cement. **Intumescent** materials react to heat by expanding and forming an insulating layer. They can be applied to steelwork as sprays or paints and have the advantage of retaining the profile of the structural element. The disadvantages of intumescents lie in the more limited length of fire resistance they can provide (normally 1 hour, as opposed to a maximum of 2 hours for sprayed coatings or 4 hours for boarding), and in the susceptibility of the material to abrasion damage either during construction or during the lifetime of the building. Intumescents must also be applied over suitably cleaned and primed steel, and the thickness of the layer must be checked before a top sealer coat is applied. (Figure 5.6).

Dissipation of the heat away from the steel is a more exotic option, but it is possible to use water to cool hollow steel sections or flame shields to reduce heat

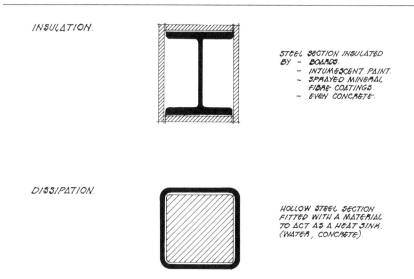

Figure 5.6 Protecting steelwork

gain. It is even possible to design the structure such that it is outside the building envelope and therefore protected from the risk of internal fires. However, these techniques are very expensive and require maintenance throughout the life of the building.

5.1.5 Concrete

It is possible to achieve very high levels of fire resistance with reinforced concrete; up to 4 hours is quite easy. However, as reinforced concrete depends for its tensile strength on the steel reinforcement, it is critical that in the design of the elements sufficient protection is provided to the steelwork. Simply increasing the thickness of the concrete cover to the reinforcement does not necessarily give a corresponding increase in safety because of the tendency of concrete to spall (break off) in a fire. This can reduce the cover and it may be necessary to provide supplementary reinforcement to counteract this danger if the cover is thicker than 40 mm. One of the critical issues in the fire resistance of concrete is the nature of the aggregate which has been used, certain aggregates being more resistant to spalling and having a lower thermal conductivity. The issue of thermal conductivity is particularly important when the assembly is also providing a subdivision and it is necessary to limit heat transfer. Also critical can be the use of permanent steel shuttering when it is necessary to design the concrete slabs to be able to withstand the failure of the steel.

5.1.6 Brick

Brickwork is generally a very good fire-resisting material; and it is quite possible to achieve periods of resistance of up to 4 hours, the stability of the material being due to the high temperatures to which it has already been subjected during manufacture. However, there may be problems with large panels (over 4 m) of brickwork due to differential expansion and movement. In these instances, the restraints being applied to the edges of the panels become critical (e.g. brick panels in concrete frame buildings).

5.1.7 Glass

Normal glass has very little fire resistance, offering little insulation and being liable to lose its integrity and stability as it shatters under fire conditions. However, there are three types of glass now marketed which offer some degree of fire resistance. The familiar Georgian-wired glass can solve the problems of stability and integrity by holding the glass in place, but this still does not offer any insulation, and radiant heat can still pass through the material. **Toughened glasses** are now available (e.g. Pyran from Schott, and Pyroswiss from Colbrand) which achieve the same integrity and stability as the wired glass without the unattractive appearance of the wires, yet these also fail to provide any insulation. The one type of glass which does offer insulating properties is **laminated glass** (e.g. Pyrostop from Pilkingtons, and Pyrobel from Glaverbel). These incorporate a completely translucent and transparent intumescent layer, which on the application of heat expands to form an insulating barrier. The disadvantages of such laminated glass lie in its weight, cost and limitations on external use. Such glass must also be ordered pre-cut, as this is a factory rather than a site job. With all three types of glass (wired, toughened and laminated), the design of the frame is as important as the choice of glazing material and it is essential that the frame will survive as long as the glass. It is crucial that the architect considers the fire resistance of the glazing assembly and not just the glazing material itself.

5.2 Passive measures: compartmentation

The compartmentation of a building into a series of fire- and smoke-tight areas will contain fire spread and gain time, the fire being contained while the occupants have a chance to escape or to take refuge until it can be extinguished. Compartmentation also offers the chance of containing the fire to protect, at least, the rest of the property while the fire is extinguished. Therefore compartmentation is important both for life safety and for property protection.

The fire protection of elements of structure not only ensures that the building does not collapse, but can also help to compartment the building. However, to achieve complete separation into different compartments it will also be necessary to protect some non-structural elements such as internal walls and doors. The fundamental principle for the designer to remember is that the integrity of the subdividing elements must be maintained and there can be no weak points or cavities which break the fire and smoke barrier. Any services or ducts which breach the compartment walls or floors must be designed to provide an equal level of fire resistance. A major threat to the fire safety of a building can come from the late addition of services or ducts which are cut through dividing walls by subcontractors unaware or uncaring that they are breaching critical fire barriers. Any small gaps of imperfections must be **fire stopped**.

Any doors through **compartment walls** must not only match the fire resistance of the walls, but trouble must also be taken to ensure that they can close swiftly in the event of a fire. The simple door-wedge poses another major threat to the fire safety of many buildings. The size of compartments, and therefore the frequency of the subdivisions, has normally been specified through building legislation, but with the move towards functional requirements rather than prescriptive regulations it is important for the designer to be able to understand the funda-mental principles upon which compartmentation is based (Figure 5.7).

The number of **compartments** into which each storey should be subdivided depends on the number of people and the amount of fuel on each level. This, in turn, depends on the function of the building, and many regulations specify a maximum floor area or cubic capacity for compartments by the building's purpose. Obviously, each storey should be divided into a minimum of two com-partments, so that horizontal escape from one to the other is always a possibility for the occupants. The more combustible the contents of the building, the smaller should be the size of the compartments. A warehouse with a high fuel load (e.g. holding paint) should obviously be divided into smaller compartments than one with a low fuel load (e.g. storing steel sections), yet most legislation takes no account of this.

It is common for every floor to be a **compartment floor** as it is fairly easy to achieve fire and smoke resistance in the floor construction. However, if each floor is to be a separate compartment, then the architect must ensure that exits from each floor to the staircases, lifts, and so on, are also given an equivalent level of fire and smoke resistance. Compartments do not have to be on a single storey, and it may be that a compartment also includes a staircase leading to a small gallery. The geometry is unimportant; what *is* important is that integrity of the compartment divisions is maintained.

The spacing of compartment walls may also be determined by the ability of the occupants to escape. The maximum acceptable travel distance may be such that it becomes the key factor in deciding where compartment divisions should be placed. The fuel load of the building might be such that two compartments on each storey would appear adequate, but three may be necessary to ensure that none of the occupants is too far away from a place of relative safety.

It is common to provide an hour's fire and smoke resistance to compartment walls and floors, but this may have to be higher where they are also structural elements. Where it has been necessary to introduce additional dividing walls

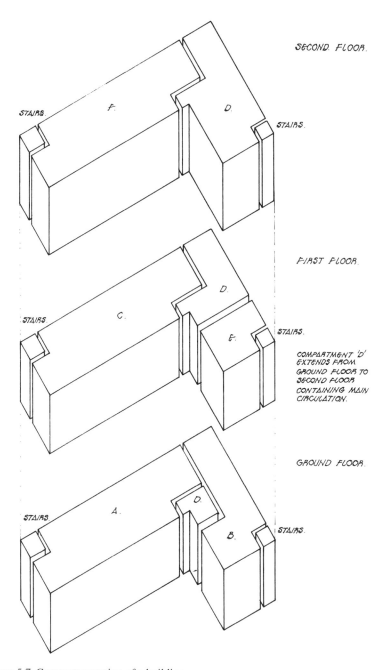

Figure 5.7 Compartmentation of a building

to reduce the travel distances these are sometimes referred to only as sub-compartments, and then only 30 min resistance is normally provided. Obviously, if the **subcompartment** walls are also structural elements, they may also have to have a higher fire resistance – and this then offers an additional factor of safety (Figure 5.8).

In addition to dividing the building up into compartments based on fuel load, the architect may also have to consider the fire protection of the escape routes from the building and treat these as additional compartments. The vertical shafts containing the stairs and lifts will need to be provided with fire and smoke resistance, and it may be necessary to isolate the routes to these shafts at the upper levels or from the foot of the shafts to the outside at ground level. These are normally referred to as '**protected shafts**' and 'protected routes', and they need the same level of fire resistance as other compartments within the building. Once an occupant of the building escaping from a fire enters such a protected route, they

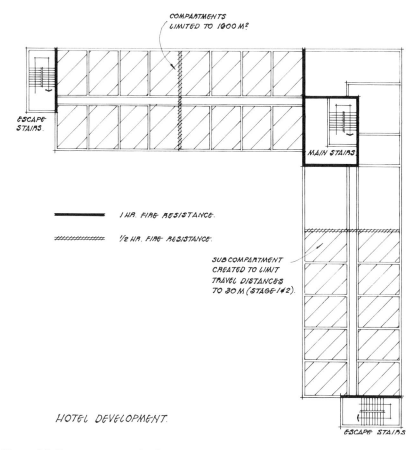

Figure 5.8 Compartments and subcompartments

Table 5.3 Building type and compartment size

Building type	Size of compartment
1 Houses	Each dwelling separate
2 Flats and maisonnettes	Each dwelling separate
3 Residential institutions (hospitals, prisons, etc.)	1600 m^2
4 Hotels and boarding-houses	2500 m^2
5 Offices and commercial	2500 m^2
6 Shops	2500 m^2
7 Assembly and recreation (theatres, cinemas, etc.)	1600 m^2
8 Industrial	
(a) high ignition hazard (oils, furniture, plastics)	900 m^2
(b) medium ignition hazard (garages, printing, textiles)	1600 m^2
(c) low ignition hazard (metal working, electrical,	
cement)	2500 m^2
9 Storage	
(a) high fuel hazard	900 m^2
(b) medium fuel hazard	1600 m^2
(c) low fuel hazard	2500 m^2
10 Car-parks	No limit

must be able to exit to the outside at ground level without encountering any further hazards.

Although each project should be assessed separately as part of a full safety engineering process, a rough guide of the maximum sizes of compartments for different building types is provided in Table 5.3. This is derived from the first principles approach of relating fuel load to building type (Table 5.1) rather than any particular code or approved document. It is based on the principle for non-domestic buildings of a maximum size of 900 m^2 (30 m × 30 m) for very high fuel load, 1600 m^2 (40 m × 40 m) for high fuel loads and 2500 m^2 (50 m × 50 m) for medium fuel loads. This is linked to the assumption that every floor will be a compartment floor. With domestic dwellings, each unit must be a separate compartment.

Table 5.3 is intended primarily for student architects working at the sketch design stage. These figures are a very rough guide and take no account of the concept of equivalency outlined in Chapter 1. It might well be that if alternative fire safety measures are included in the design (e.g. auto-suppression), then the size of a compartment may perhaps be safely increased. The special risks associated with vary tall buildings (over 10 storeys) or deep basements (more than one level) would need special attention.

We have already mentioned in the previous section the fire resistance of the major building materials from which compartment walls and floors are likely to be constructed, but doors require special consideration. All compartments are going to be breached by door openings and it is therefore critical that what is used to block these openings in the event of a fire achieves the same level of fire and smoke resistance as the rest of the wall (Figure 5.9). The term '**fire door**'

Figure 5.9 Relative compartment sizes and escape distances

is now so abused that its use without qualification is almost dangerous. The architect must know precisely the level of fire and smoke resistance offered by the full doorset, the frame and ironmongery being as important as the door leaf itself. The architect needs to know for how long the doorset will function as a barrier to heat and smoke and this should be specified in minutes. It is now possible to obtain doors which will provide in excess of 90 min fire resistance, even with two leaves opening through 180°. It is always better if trying to provide high fire resistances of 60–90 min to achieve this through two sets of doors, each having half the fire resistance, rather than a single set. There is always the danger of doors being left open, or having to be opened for escape, so a double set increases the safety factors, by providing an 'airlock'.

The designer must obtain guidance on the type of frame and the appropriate ironmongery to obtain the desired fire resistance whether or not doorsets are being used. Doors will also need suitable devices to ensure they remain shut, or that they will close immediately in the event of a fire. There are many ways of achieving this, but the architect must be careful to ensure that the chosen method is suitable for the use of the building. It may also be necessary for the door to be used for escape, so care must be taken in the choice of locking devices.

5.3 Passive measures: envelope protection

The third role of passive fire resistance is to limit the threat posed by a fire to adjoining properties and people outside the building, and to limit the possibility of ignition by a fire in an adjoining property. In this, it is the roof and the external walls that require the architect's attention, the roof because of fire spread by convection currents and the external walls because of radiant heat.

The roof can prove a danger because, once well alight, flaming particles (timbers, etc.) might be carried upwards by the convection currents and pose a hazard if they land on other buildings, these are often described as '**burning brands**'. Standards exist for designing roof constructions to resist penetration and fire spread when subjected to flame and radiant heat. However, there are no tests concerned with limiting the ability of the roof to burn to produce burning brands. It is possible to design your building to resist the threat from others, but harder to design your building to prevent its posing a threat (Figures 5.10 and 5.11).

The external walls need careful consideration as heat radiated through them from the burning building might ignite adjoining buildings if they are too close. The traditional way of limiting the danger of radiant heat is to restrict the number of openings in the external walls if it is close to other buildings. Many regulations incorporate complex calculations relating building purpose and distance from the boundary to the amount of **unprotected areas** which will be permitted. Designers should at least be aware of the use of surrounding buildings and plan both to protect their own building from outside threats and to minimize the threat which it itself poses.

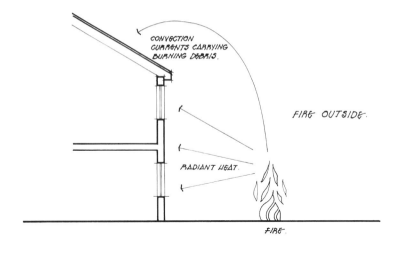

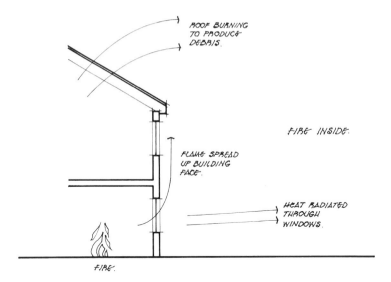

Figure 5.10 Envelope protection

There is also, of course, the danger of fire spread up the face of the building, and it is important to attempt to minimize the possibility of this happening. This can be done by careful selection of facing and roofing materials, and it is fortunate that most of those commonly used (e.g. brick, stone and concrete) have a zero fire spread potential. Some building regulations limit the use of external facing materials with a high fire spread potential in situations close to the boundary with another property. The limitation of the size of openings to reduce radiant

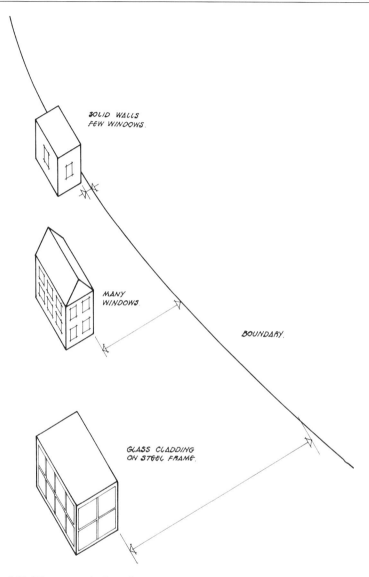

Figure 5.11 Distance to the boundary

heat to adjoining properties will also help to reduce fire spread from floor to floor, although in a well-developed fire this is hard to eliminate completely unless there are no windows at all.

5.4 Active measures

The three forms of fire containment so far considered have all been passive, in that they are properties of the building's construction which will serve to limit the spread of fire and smoke whenever a fire should occur. In addition to these passive measures, it is possible for designers to incorporate active measures of containment into their schemes – measures which will operate only in the event of a fire. Such active measures are mostly concerned with the particular problems of smoke control and the limitation of the spread of smoke throughout the building.

The toxic and lethal qualities of smoke have already been considered, and it is most important for life safety to ensure that people and smoke are kept apart. Most smoke control systems are designed to keep smoke away from the escape routes; however, some systems also assist directly in fire extinguishment. This can be done by keeping the approaches to a fire free of smoke, so that the fire-fighters can tackle the seat of the fire more safely and quickly. Smoke control systems can also help to reduce the heat damage on the structure of a building by releasing the hot gases. However, smoke control is not as effective a method of reducing heat damage as auto-suppression because the inlet air needed to balance the release of hot gases will help the fire to grow. This section is principally concerned with the protection of people from smoke, but any methods of active containment will also help firefighting and, to some extent, reduce heat damage. There are two principal methods of active containment of smoke, **pressurization** and **venting**; these will now be considered in turn.

5.4.1 Pressurization

We have already mentioned the problem of designing doors for fire-resisting walls such that they achieve a level of fire and smoke resistance commensurate with the surrounding walls. Even when well designed, it is inevitable that doors on escape routes will have to be opened and that smoke will therefore flow into the protected area. This danger can be reduced by using lobby access to staircases, which provides a form of 'airlock' where (hopefully) only one door will be open at any time. However, this is far from ideal, and a much better way to resist the inflow of smoke is to pressurize such protected areas, either corridors or stairs. It would also be possible to attempt to keep the escape route smoke-free by extracting smoke as it enters, but this will tend to draw more smoke into the area. Therefore smoke ventilation is more appropriate to large spaces rather than confined staircases or corridors, and this is considered separately in section 5.4.2. Pressurization is more suitable for spaces where the volume of air is smaller, and it is possible to consider raising the air pressure, so that smoke can be prevented from flowing in. Pressurization is not only used in fire situations, but also in areas where it is important to maintain a 'clean environment' and keep out contaminates of any form, for example, in operating theatres or in factories assembling electronic equipment (Figure 5.12).

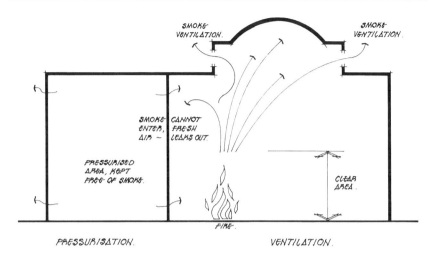

Figure 5.12 Pressurization and ventilation

Fresh air is supplied to the area to be kept smoke-free, and the air pressure is maintained at a level above the surrounding rooms. Therefore, if a door into the pressurized area is opened, air will flow out rather than smoke flowing in. When all doors are kept closed, the positive pressure which has been built up will prevent any leakage through cracks into the area and the fresh air will leak out into the adjoining spaces.

The amount of air which must be supplied will be determined by the air-leakage characteristics of building construction, the likely number of open doors to the protected area (assume a minimum of one in 20) and the other pressures acting on and in the building which will influence airflow patterns (e.g. the 'stack effect'). It is interesting to note that the volume of space being protected does not enter into the calculations, except in so far as it is related to the leakage characteristics of the building.

Provision must be made such that air which has passed from the pressurized protected areas to the unpressurized spaces can be vented to the outside. This is crucial to maintain the pressure differential which ensures the effectiveness of the system; this may well be adequately provided by the windows on each floor – but it may be necessary to consider additional vents or even mechanical extraction of the air (Figure 5.13).

A pressurization system can either be designed to act only in the event of a fire (single stage) or it may act continuously at a low level and then increase in air supply when a fire is detected (two stage). The two-stage system is preferable because it always gives some measure of protection and may limit the earliest stages of fire spread, before detection even occurs. Care must always be taken that the effectiveness of the pressurization system is not in any way jeopardized by other air-handling systems within the building.

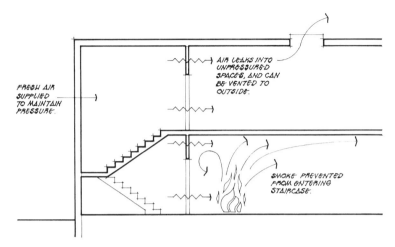

Figure 5.13 Pressurization

It is possible to pressurize just the staircases, but it is obviously much better (and leakage will be substantially reduced) if the staircase lobbies are also pressurized. The ideal solution is to pressurize the whole escape route, including the horizontal in addition to the vertical parts. A separate system should be designed for each staircase to prevent failure on one staircase affecting the others.

The distribution of the air to the protected areas must be such that the protected area is pressurized evenly. Therefore in staircases the supply air must be ducted for the height of the enclosure with inlet grilles at intervals not greater than three storeys. A single inlet would not be acceptable on any building over three storeys in height. For pressurization to be successful, the supply of fresh air being forced into the protected area must be constantly maintained through the life of the building. The air supply must be reliable and intakes properly maintained.

5.4.2 Venting

The simplest way of stopping smoke spread within a building is allowing the smoke to escape to the outside. While this will not extinguish the fire, it can contain the smoke to its area of origin and gain time for people to escape and for measures to be taken to extinguish the fire. In a single-storey building this can easily be done through roof vents, but it is also possible to design smoke ventilation systems for multi-storey buildings using mechanical extraction.

The first essential for the designer to understand is the different zones which will develop within the smoke. The hot, smoky gases from the fire will form the upper stratified layer below the ceiling (Zone A). They will float on the colder smoke-free air below (Zone B). The plume of smoke ascending from the fire entrains air as it rises and creates the upper layer (Zone C). This **stratification**,

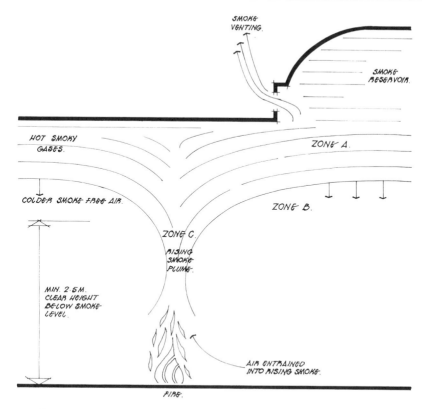

Figure 5.14 Ventilation

or **layering**, of smoke is due entirely to the buoyancy of the smoke being produced, and should the smoke cool, then the stratification will break down (Figure 5.14).

Smoke production will increase exponentially with the growth of the fire. Although it may be possible to assume that initially the smoke will be able to exit directly through the vents in the roof, as the fire grows a layer of smoke will build up beneath the ceiling. The layer of smoke will get thicker as the fire grows, and the smoke level will gradually descend. The increasing depth of the smoke layer will increase the pressure on the vents that are available – expelling more smoke – and reduce the capability for the fire to entrain more air, reducing volume of smoke produced. The venting system must be designed to ensure that the smoke being added to the smoke layer is exactly balanced by that being expelled through the vents, so that the depth of the smoke layer remains constant and never descends to a level where it endangers the occupants. The objective will be to have at least 2.5 m clear height beneath the smoke level.

Limitation of the sideways spread of the smoke can be achieved by the installation of '**smoke curtains**', barriers which come down from the ceiling and create 'smoke reservoirs'. Such smoke curtains might be permanently in place or be triggered to fall by the fire. Smoke reservoirs are desirable because they limit the area of damage caused by the hot, smoky gases, and ensure that the vents can operate to their maximum efficiency (Figure 5.15).

The screens which are used to contain the smoke should ideally be as fire resistant as the roof structure itself. The use of different depths of screens in different directions can ensure that if smoke production is too great for the capacity of the reservoir, then the spillover will occur into the less rather than more dangerous areas (Figure 5.16).

Smoke reservoirs need not necessarily be formed by the downstand of curtains; they can equally well be formed by upstand areas into which smoke can flow. The high vaulted roofs of a design can often be utilized in this way, so that

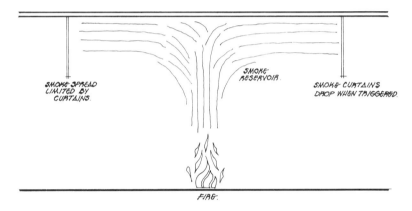

Figure 5.15 Smoke reservoirs

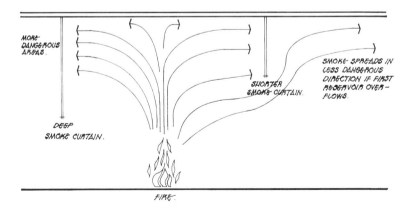

Figure 5.16 Smoke curtains of different depths

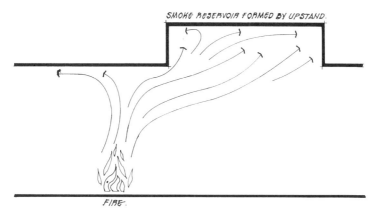

Figure 5.17 Use of upstands

smoke is contained above head height until it can be expelled from the building (Figure 5.17). It is quite possible to design the size of the smoke reservoirs and the capacity of the ventilation system to ensure that the smoke level does not descend below a dangerous level. This would not normally be the responsibility of the architect, but it is necessary for the architect to understand the principles which would underlie such a design.

The first principle which the architect needs to be aware of is that of fire size. It is essential to recognize the assumptions which have to be made about the size of fire that is likely to occur. Sprinkler systems are normally designed to limit a fire to a 9 m² area (a 12 m perimeter). In retail premises it has been estimated that this would lead to approximately a 5 MW fire, assuming 0.5 MW/m² of fuel load. However if fast response sprinklers are used the fire may be restricted to 5 m² giving a fire design size of 2.5 MW. In calculating smoke production from fires in sprinklered retail buildings, it is assumed that these figures represent the largest probable fire. In other sprinklered buildings it may be appropriate to use the 5 MW or 2.5 MW fire for design purposes, but care should always be taken to consider the specific nature of the ignition risk and the fuel load. In unsprinklered buildings, where combustible material is arranged in stacks with aisles or spaces between, then the fire size will be taken as the largest of these stacks. In other unsprinklered premises the estimation of fire size becomes much more difficult and an estimate must be made based on the fuel load and disposition.

Having determined the maximum size of fire to be considered, the architects need to establish how low they can permit the base of the smoke layer to descend. This will normally be at least 2.5 m above the floor level, so that escape routes are not jeopardized. Once the height of the base of the smoke layer is established, then the size of the available smoke reservoir can be calculated from the geometry of the design. This must be large enough to store the smoke produced until it can be extracted. The smoke reservoirs cannot be infinitely large because of the dangers of the smoke cooling as it comes in contact with the clear air below. Therefore the limiting factor on reservoir size is normally the plan area. For high-risk buildings (e.g. shopping centres) the maximum reservoir size should not exceed 1000 m³, and for low-risk buildings (e.g. sports halls) perhaps 2000 m³. These figures are extremely crude guides for architects; the detailed design of the smoke ventilation system is a matter for specialists.

The architects must be aware of the relative advantages of mechanical and natural ventilation in different situations (Figure 5.18). Mechanical ventilation must be designed with reliability in mind, so that it can be repaired and maintained throughout the life of the building. Natural ventilation is totally responsive to the ambient conditions, and the external weather conditions will determine the effectiveness of the ventilation system. If the wind pressure on the outside of a vent is greater than the smoke pressure on the inside, then fresh air might flow in through the vent, cooling the smoke and forcing it down into the building. If the vent is in a flat roof, then the effect of wind blowing across the opening will be a suction effect which will improve the efficiency of the vents. However, if the vents are in a pitched roof, and particularly a steeply pitched roof, those on the windward side might not function and could even be counterproductive. Vents in this position need to be specially designed to overcome this problem.

Any decision to use natural ventilation will be dependent upon the site conditions, surrounding buildings, potential for new tall buildings, and the prevailing winds.

The final principle for the designer to be aware of concerns the provision of inlet air to replace the ventilated smoke. If no provision is made to allow fresh

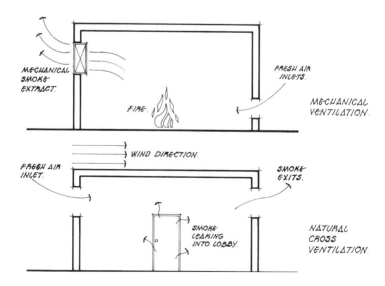

Figure 5.18 Mechanical and natural ventilation

air to be drawn in to replace that leaving through the vents, then the system will fail to operate. It is preferable if these inlets are widely distributed over the building, so that it does not matter where the fire occurs. The use of the doors of a building to act as the inlets may be possible, but provision would have to be made to ensure that the doors opened automatically in the event of fire.

At some level within a building filling with smoke, there will be what is called the **neutral plane** (Figure 5.19). The inlet air will be drawn in below this because the air pressure within the building is below atmospheric pressure, while air pressure above the neutral plane is above atmospheric pressure and the smoke is being forced out through the vents. It is important in the design of the smoke venting system that this neutral plane is high enough to limit smoke spread within the building.

In a building with a sophisticated air conditioning system, it is essential that attention is paid to the potential impact of the system on the smoke generated by a fire. If well designed, the system could be used to vent smoke from part of the building and to supply the necessary balancing inlet air. However, this is far from easy to design and control, and in most fire situations the air conditioning plant is normally designed simply to shut down or is set only to extract air. Badly designed equipment could spread the fire very rapidly by re-circulating smoke. Whatever system is used, manual override must always be provided, so that the fire service can take charge once they arrive.

The relationship between smoke ventilation and the operation of sprinklers is a complex one, but the architect needs to be aware of the fundamental principles. The provision of sprinklers should reduce the risk of the fire growing beyond

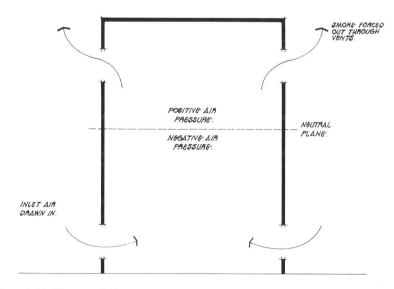

Figure 5.19 The neutral plane

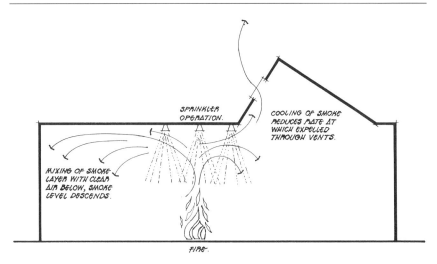

Figure 5.20 Sprinkler operation and smoke ventilation

certain limits – and these limits are used as the basis for the design of the smoke extract system. However, the water spray from the sprinklers might cause mixing of the smoke layer with the clear air below and bring down the smoke level. The sprinklers might also cause cooling of the smoke, which will reduce the rate at which it is being expelled through the vents. However, these possible dangers due to the cooling and mixing of the smoke can be counteracted by increasing the smoke vent size to ensure that the smoke level does not drop, and that smoke will still be vented even if the temperature is slightly reduced (Figure 5.20).

Not only may the sprinklers have a detrimental effect on the operation of the vents, it is also possible that if the vents open before the sprinklers operate, then their operation may be slightly delayed. However, in most instances, this delay is now thought to be negligible. The complexity of the relationship between sprinkler operation and smoke venting should be known to the designers and they should be conscious that the proper design of such a system may well require expert help.

All types of containment, whether of heat or smoke, and whether active or passive, only gain time. They provide the opportunity for the occupants to escape and for an attempt to be made to extinguish the fire. Fire extinguishment is the final tactic available to the designer and will be considered in Chapter 6.

Extinguishment 6

However effective containment is in limiting the spread of a fire, eventually the fire must be extinguished. In Chapter 1 the 'triangle of fire' was described, and it was stressed that the removal of any of the three elements (heat, fuel or oxygen) would terminate the chemical reaction, so extinguishing the fire. In the open a fire may be left to burn itself out (uses up all the fuel), but in a building (even after the occupants have escaped) the fire must be extinguished to limit the damage to property. The most common extinguishing agents are water, foam, carbon dioxide and dry powder (Table 6.1).

Water is the most common extinguishing medium and works by cooling the fuel (removes the heat). To a lesser extent, water will also act as a smothering agent (removal of oxygen). Unfortunately, water is a very effective conductor of electricity and so cannot be used on electrical fires. The other disadvantage of water is that it does not mix with oil-based products and so cannot be used for many liquid fires. Water extinguishers have traditionally been colour-coded signal red.

Foam works both by cooling and smothering, and it is excellent for extinguishing certain liquid fires. There are various types of foam with differing levels of expansion. High-expansion foam is used in some special installations and by fire brigades to completely fill an area. Low- and medium-expansion foams are used to provide a covering blanket of foam over the fire. Foam extinguishers are normally colour-coded pale cream.

Table 6.1 The suitability of extinguishing medium for different fires

	Fires in solids	*Electrical fires*	*Fires in liquids*	*Fires in gases*
Water	Excellent	No	No	No
Foam	Good	No	Excellent	No
CO_2	Poor	Good	Good	Good
Powder	Poor	Excellent	Excellent	Excellent

Carbon dioxide will smother a fire and is particularly good on electrical fires as it does not act as a conductor. A concentration of approx. 25–30% of carbon dioxide in air is necessary to extinguish a fire. Carbon dioxide is toxic and a concentration of 12% in air can be lethal; it is best used in areas which have been evacuated, or inside equipment cabinets. Very occasionally, nitrogen gas is used instead of carbon dioxide as a smothering agent. Carbon dioxide extinguishers are normally colour-coded black.

Dry powder works by suppressing the combustion reactions in the flame. There are a number of different powders commercially available, including sodium bicarbonate (baking soda). Dry powder extinguishers are normally colour-coded French blue.

The halogenated hydrocarbons (**halons**) have been used in the past as extinguishing agents as they are even more efficient anti-catalysts and if supplied in the correct concentrations can extinguish a fire almost immediately. The two halons most commonly used were BCF or Halon 1211 (bromochlorodi-fluoromethane) and BTM or Halon 1301 (bromotrifluoromethane). These are gases at room temperature, but are stored as liquids under pressure. Halons can be used on any type of fire and, unlike the others, once the gas is dispersed, there is no additional damage from the extinguishing medium. A concentration of approx. 3–6% of halon in air (by volume) is necessary to extinguish a fire.

Unfortunately, halons had two very significant drawbacks. First, they are toxic – a concentration of 10% of halon in air (by volume) can be lethal; therefore they could only be safely used, like carbon dioxide, in uninhabited areas. The toxicity is associated with the decomposition products formed as the halon is exposed to the flame. The use of manual extinguishers is safe, provided that there is adequate ventilation; however, at least one fatality has occurred after halons were used to extinguish a soldier's clothes inside a closed vehicle although the burns alone would not have been fatal. The second drawback concerns the damage their long-term use will cause to the earth's ozone layer. These problems mean that the use of halons is now restricted to very specialist situations. Ironically, halon extinguishers are normally colour-coded emerald green.

There are three principal methods of applying the extinguishing agents: first, by the occupants themselves with manual firefighting equipment; second, through auto-suppression systems; and finally, by the fire service. Architects and designers must consider during the design stages which of these methods they expect to be used and then design to ensure their effectiveness.

6.1 Manual firefighting

The importance of manual firefighting equipment provided for use by the occupants should be recognized by architects and designers. In industrial situations approx. 90% of all fires are extinguished by people using hand extinguishers or hosereels. The architect can play a role in ensuring that the equipment provided is of the right type, in the right numbers and in the right places. It is the responsibility of the building owners to ensure that the staff are adequately trained to be able to use the equipment. Equipment can be divided into three categories: hand-held extinguishers, fire blankets and hosereels.

6.1.1 Hand-held equipment

The number of hand-held extinguishers to be provided will depend upon the size of the building and its use. If the building is divided into compartments, then the absolute minimum provision will be two water extinguishers in each compartment. Most extinguishers will be water, but other types must be provided for special risks (for example, carbon dioxide for each electrical switch-room or plant-room). The extinguishers should be positioned so as to be as accessible as possible. It is normal to site them near to the entrances to each compartment, so that they are on the escape routes and also available to staff entering the area to fight the fire.

Extinguishers have suffered in the past from a variety of methods of operation and colour-coding systems. Colour codes have now been standardized and the extinguishing medium is indicated either by a broad band of the relevant colour on a red or metal extinguisher, or the whole extinguisher is colour-coded. Most extinguishers now operate on the removal of a pin and then by squeezing the trigger.

Training is very important if people are to use hand-held extinguishers effectively. They have a fairly short discharge time (normally less than 2 min) and unless used to maximum effect will run out before they have any impact on the fire.

6.1.2 Fire blankets

These are useful in extinguishing a fire by smothering and are particularly valuable in kitchens where there is a danger of fat fires. They can also be used for wrapping round someone whose clothes are alight. They used to be made of asbestos or leather, but are now normally made from fibreglass.

6.1.3 Hosereels

Hosereels provide a larger supply of water than hand held-extinguishers, but water is not suitable for all types of fires. They are no harder for the occupants to use than hand-held extinguishers (particularly if trained) and are often also used by the fire service. Unfortunately, they are much more expensive to install and maintain.

There is a practical limit of about 30 m to the length of the hose, and with a jet of about 6 m, this means they should really be located so that no place in the building is more than 36 m from the hosereel. They should be located such that they can be used within a compartment rather than having to be taken through compartment doors, breaching the integrity of the barrier. While the architect will want to design the hosereel housings such that they do not dominate the building, if too cleverly concealed, the occupants might not realize what they are. The other difficulty with built-in hosereels is gaining access for regular maintenance and checking.

6.2 Auto-suppression

Auto-suppression systems are those which are activated in the event of a fire without any action by the occupants. The most commonly installed form of auto-suppression is sprinkler protection, but other forms are available for special risks.

6.2.1 Sprinklers

Water sprinklers have been used since the end of the nineteenth century and they are considered so effective in minimizing property losses that insurance companies will give substantial discounts (up to 60%) on the premiums to building owners who have installed them.

Sprinkler systems are designed to extinguish small fires or to contain growing fires until the fire service arrives. Sprinkler heads are heat-sensitive, and it is possible to decide the temperature at which they should activate. Normally this is 68 °C, but in certain situations it may be necessary to be set at a higher level. It should always be designed to be at least 30 °C above the highest anticipated temperature. Each sprinkler head acts as its own heat detector and only those in the fire area will be activated. The maximum area to be covered by each head depends on the risk, but 9 m² is normally taken for all but the most special risks in high hazard areas. This gives the maximum fire size already mentioned for smoke calculations, and it is expected that the fire will be contained within this area (3 m × 3 m, giving a 12 m perimeter). If fast response time sprinklers are used then the fire may be contained to a smaller area, perhaps half this size. Sprinklers will not be able to extinguish a fully developed fire as the operation of too many heads would reduce the water pressure at the open heads to such an extent that it would be insufficient to suppress or control the fire.

The contribution of sprinklers to life safety is harder to quantify, and their value lies in the limitation of fire while the occupants have an opportunity to escape. However, issues of the interaction of sprinklers and smoke ventilation, mentioned at the end of Chapter 5, must be considered by the designers.

The detailed design of a sprinkler system is certainly not a job for the architects; however, they should be aware of the different systems available. '**Wet**' **systems** are full of water at all times, while in '**dry**' **systems** the majority of the pipework inside the building is air-filled until it is triggered (thereby minimizing the risk of freezing in cold weather). '**Alternating**' **systems** can be changed from 'wet' operation (in summer) to 'dry' operation (in winter).

'**Re-cycling**' or 're-setting' sprinklers are ones where the heads can be closed, once the fire is extinguished (minimizing water damage). A '**pre-action**' **system** is one which is 'dry', but where water is allowed in on a signal from a fast-response detector (usually smoke) in advance of the heads being triggered.

Sprinkler heads do not always have to be ceiling-mounted, they can be wall-mounted, and in the case of aircraft hangers they have even been installed at floor level to spray the undersides of aircraft. In high bay warehouses it may be desirable to have sprinklers at different levels within the stacks as ceiling-mounted sprinklers could not reach the seat of the fire.

Normal sprinklers are not as responsive as either heat or smoke detectors and might easily take twice as long to activate. To counter this lack of sensitivity, 'fast-response' or 'residential' sprinklers have been developed. The better sensitivity is achieved by improvements to the heat-collecting arrangements and a reduction in the thermal lag of the system by using lighter components. 'Fast-response' sprinklers can be as sensitive as heat detectors.

The architect does need to ensure that the sprinkler designers have precise information about the different ignition risks and fuel loads in each compartment. They will also have to co-ordinate the sprinklers with other service installations and settle the positions of the main stop valves and, perhaps, alternative water supplies. To ensure that the sprinklers have an adequate water supply two separate sources are sometimes required. One of these may be the normal mains, the other might be private reservoirs or storage tanks.

6.2.2 Other forms of auto-suppression

The decision on the installation of more specialized forms of auto-suppression will depend on the ignition risks, the fuel loads and the nature of the property to be protected. It will also be dictated by cost as they are all considerably more expensive. If a system is to be installed, it will probably be to attract an insurance premium discount, and therefore the standards of the insurers will have to be met by the architects.

In addition to standard sprinklers, water can also be used in other forms of auto-suppression. Water spray systems are designed to extinguish flammable liquid fires or cool tanks and are more applicable in industrial plant than in buildings. Drenchers or deluge systems are intended to protect the external faces of a building or a compartment (walls, windows and roofs) from damage by exposure to fire in adjacent premises. These may be manually triggered or linked into the fire detection system. Drenchers can also be used to cool elements of structure (for example, steel roof beams), and they are particularly valuable as supplementary protection over openings within compartment walls.

It is quite possible to use carbon dioxide as an auto-suppression system. It has the advantage of doing little damage to the contents of the compartment and of being usable in electrical fires. However, the toxicity of carbon dioxide limits its use to areas which are unoccupied, or to the protection of individual items of plant (for example, computer cabinets). It can be used for areas where people occasionally go, provided that there is sufficient warning of the discharge of the system to enable the occupants to escape. A particular use might be to protect the storerooms of a gallery or museum.

Foam can be used to protect particular items of plant or, in the case of high-expansion foam, completely flooding a compartment. As foam will be fairly damaging to the contents, it is normally used only for heavy plant or machinery rooms. It is also possible to install auto-suppression systems using dry powder as the medium. As with foam, the powder is damaging to the contents of the compartment, and it is only in very special situations where this becomes the preferred option.

6.2.3 Equivalency of auto-suppression

The decision on whether or not to include some form of auto-suppression system within the design may well be determined for the designer by the client's insurance company. However, if the architects are adopting the fire safety engineering approach to the design, they will also wish to consider the additional safety offered by such systems and compare it with alternative methods of achieving equivalent safety, perhaps, through compartmentation. In a fire safety engineering approach to a complex building a more detailed analysis of equivalency would be necessary, including the role of smoke ventilation, the communications and, most important, the ignition risk and fuel loads. This is an area of specialization where the designers will probably use an outside fire safety consultant. However, the architects do need to be able to understand the concepts put forward by their consultants and to ensure that they are satisfactorily integrated into the total design of the building.

6.3 Fire service facilities and access

6.3.1 Facilities

In tall buildings, where the fire must be tackled from within, the fire service require safe '**bridgeheads**' from which they can work (Figure 6.1). These need to be linked by specially protected lifts which, in the event of a fire, are solely for use of the fire service. Such lifts require two independent power supplies with fire protection to the cabling. They also need good communications with ground level and they must be fully under the control of the people within the lift.

As the basis of the 'bridgeheads', the fire service need access to **rising mains**. These are vertical pipes within the building with a fire service connection or booster pump at the lower end and outlets at different levels within the building. **Dry risers** are recommended for all buildings of over 18 m and for buildings with more than 600 m^2 at 7.5 m or over. In measuring the height it is acceptable to measure from the fire brigade access level. Dry risers are kept empty of water and the fire service

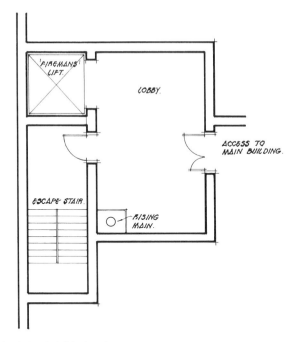

Figure 6.1 Fire brigade bridgeheads

supply water to the inlet at ground level. The fire service are then able to attack fires on upper levels without having to lay hose up through the building.

Wet risers are normally installed in taller buildings (over 6 m) and they are kept permanently charged with water. They permit the fire service to attack fires from within the building without having to provide or pump up water. Supplementary pumps may be necessary to raise the water if the mains pressure is insufficient to cope with the height of the building.

Certain plant-rooms or tank-rooms may be provided with inlets to enable the service to flood them with foam from outside the building. Appliances need to be able to get within 18 m of such inlets.

6.3.2 Access

It is essential that the architect considers fire service access at the early stages of the design as it might well influence the location of the building on the site. In the case of buildings without rising mains, and therefore totally dependent upon the hoses and equipment carried by fire service appliances, the degree of access is determined by the size of the building and the height of the highest floor level above the fire brigade access level. If the building is over 11 m high, these requirements are more stringent as hydraulic platforms or turntable ladders may have to be used to effect rescues (Figure 6.2 and Table 6.2).

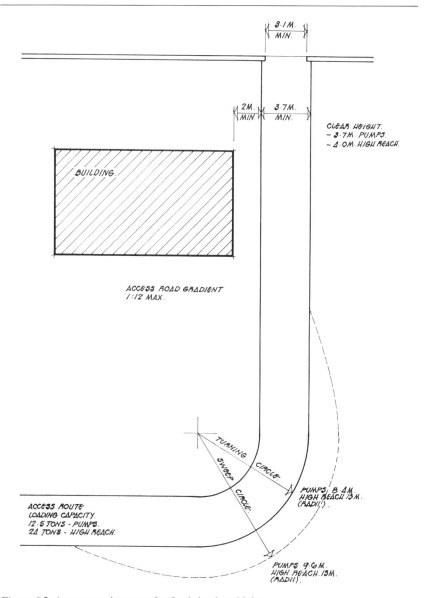

Figure 6.2 Access requirements for fire brigade vehicles

If wet or dry rising mains are provided, then provision must be made to bring pumps within 18 m of the ground-floor access point of the inlet to the main and within sight of the building. A fire safe design should therefore allow for a pumping appliance to draw up in this position and to have an adequate supply of water (i.e. hydrants) within a reasonable distance. The firefighters should not

Table 6.2 Fire brigade requirements for buildings without rising mains, giving the percentage of the perimeter to which vehicle access must be provided

| Height of building | *Total floor area of building (m²)* | | | | | Access for vehicles |
	up to 2000	*2000– 8000*	*8000– 16 000*	*16 000– 24 000*	*Over 24 000*	
Up to 11m	15%	15%	50%	75%	100%	Pumps only
Over 11m	15%	50%	50%	75%	100%	Pumps and high reach

High reach vehicles are either hydraulic platforms or turntable ladders.

have to travel far into the building, before reaching stairs (and, if provided, the fireman's lift). In excessively high buildings it may be necessary to introduce a 'fire floor' which is designed as a firefighting platform and is open to the air, with provision made for water drainage and isolation of staircases. Access roads must be capable of bearing the weight of fire appliances and provide sufficient room for them to manoeuvre.

Assessment

7

When a design project involves the refurbishment, upgrading or renovation of a building, the architect will have to assess the fire safety of what exists and determine if it achieves an acceptable level. Only rarely do architects have the luxury of a totally free hand to design a completely new building on a green field site; the majority of projects involve at least some existing buildings. Therefore the designer needs to understand how to assess such existing buildings. If it is a renovation project then it is likely that the existing building will be empty and only the physical attributes (the design issues) will have to be considered, but if it is only refurbishment then the building may well be occupied and any assessment will also have to include management issues.

With increasing concern about health and safety issues, fire safety legislation has also become more concerned with the activities within buildings and with the management of buildings in use. Risk management and risk assessment have now become key aspects of health and safety provision, and fire safety is being seen as an essential component of such analysis. At the same time the importance of the fire assessment of buildings has also grown as fire safety legislation became less prescriptive and more performance based. Traditionally the Building Regulations set standards for new buildings, but with the introduction of the Fire Precautions Act in the 1970s, standards were applied to all buildings within certain categories, and this led to assessment and improvement of existing buildings. Now the advent of European directives on workplace safety, particularly the Framework Directive (89/391/EEC) and the Workplace Directive (89/654/EEC), has meant that all workplaces have to be assessed, and where necessary improved to the acceptable standard.

Therefore fire assessment is not only relevant where general improvements were being undertaken; increasingly, fire assessment is leading to improvement works on buildings where work would otherwise not be necessary. In these cases the building will definitely be in use, and the assessment will have to take into account both design issues and, equally important, management issues.

The first section of this chapter will look at the principles which must lie behind any assessment. The next four sections will then consider a series of assessment methods which have been developed by the authors for specific building types. The penultimate section of this chapter examines the related issue of fire audits, and the final section considers the problem of finding competent fire assessors.

7.1 Methodology

Outside the academic press there is little published on the principles of fire assessments, and few workable fire assessment schemes for specific building types. However, there are a number of fundamental concepts which should underlie all such assessments. An architect faced with undertaking an assessment or negotiating with a consultant who has undertaken such an assessment needs to be aware of these and of their significance. When dealing with the statutory authorities (fire brigade, building control) then the ability to be able to work from first principles and query any unfounded assumptions or dubious reasoning becomes critical. In very large buildings the architect will probably have the backing of a specialist fire engineer, but in most medium scale jobs or improvement programmes it may be necessary to argue the case in person.

Before beginning any assessment it is essential to be clear about the terminology being used. In fire assessment the correct use of words and phrases is very important, and it is necessary to distinguish carefully between the different terms. The word '**assessment**' itself can be used interchangeably with '**appraisal**', but needs to be clearly distinguished from '**audit**', which will be discussed in section 7.6.

A fire assessment will always be in three parts: firstly, an estimate of the fire risks; secondly, an estimate of the fire precautions; and finally, consideration of the extent to which the precautions balance the risks and how the building compares with any established benchmarks (Figure 7.1). If only the precautions are considered, or only the risks, then you do not have a complete assessment. The value of fire assessment lies in the calculation of the extent to which the fire risks which have been estimated are mitigated, or compensated for, by the fire precautions which have been estimated. The fire assessment must provide usable results and not simply a listing of building characteristics; in simple terms it must tell you whether or not the building is acceptable.

A glossary of relevant fire terms is included near the end of the next chapter (section 8.7) in which definitions are given for all the most commonly used words and phrases. The following three subsections consider the three parts common to all assessments.

7.1.1 Risks

The first part of an assessment is the estimation of the fire risks within a building. This fire risk assessment process should be very similar to any other risk assessment and must follow three standard stages. Firstly, the hazards have

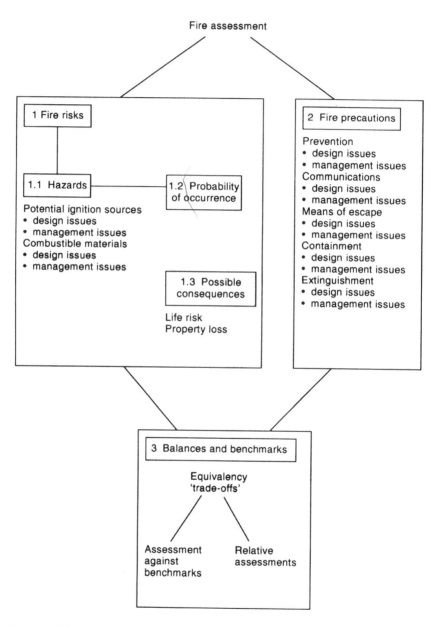

Figure 7.1 Fire assessment

to be identified; secondly, an estimate must be made of the likelihood of these hazardous events occurring (i.e. their probability); and finally, the possible consequent life and property loss has to be considered. A risk assessment therefore reflects both the likelihood that harm will occur and a measure of its severity.

Hazards are sets of conditions in the operation of a product or system with the potential for initiating a fire. They include not only consideration of the combustible materials which might burn, but also the sources of potential ignition. However, it is not sufficient just to list the hazards identified in a building; the assessor must make a judgement about the likelihood of these hazardous events occurring. For example, a soft furnishings store in a building should always be identified as a hazard because of the combustibility of the materials involved; however, the risk posed by this store will depend on the probability of ignition. In some uses the risk will be very small because sensible storage procedures have been adopted, and the staff have been trained in the handling of such materials. In other buildings a store of similar size will constitute a major risk because half the staff are unaware of its combustible nature, and no one has responsibility for ensuring that it is properly supervised. The same hazard can pose very different risks depending on how it is used.

The risks which may be identified can be related to either the design of a building or to its management. Typical design issues would include electrical installations and surface finishes. Management issues might include smoking/non-smoking policies and work processes. Some issues, such as the siting and management of a soft furnishings store, would combine both design and management issues.

The third stage in the process of risk assessment is consideration of the threat to life and property. Here the consequences of ignition to people and property have to be considered. In the example of the soft furnishings store it can be seen that risk will vary according not only to how it is managed, but also to where it is located. If it was in a hospital basement beneath a ward block, then the risk to life and property would be much higher than if it were in an isolated building on the far side of the car-park.

7.1.2 Precautions

Having assessed the risks the second part of the assessment is to examine the value of the fire precautions within the building. The existing precautions must be identified, and an estimate made of how far these will reduce the likelihood of ignition occurring and/or mitigate the consequences should ignition occur. Fire precautions can be grouped under the five fire safety tactics already covered in the previous chapters: prevention, communication, escape, containment and extinguishment.

The precautions which may be identified can be related to either the design of a building or to its management. Typical design issues would include travel distances and sprinklers. Management issues might include staff fire drills and maintenance of firefighting systems. Some issues, such as the planning of a communications system, would combine both design and management issues.

The level of detail in a fire assessment should be broadly proportionate to the risks involved. The purpose of an assessment is not to catalogue every trivial hazard and deficiency in precautions. It is also unfair to expect an assessor to anticipate hazards beyond the limits of current knowledge. The term often used in regulations is that the assessment should be 'suitable and sufficient'. This means that it must reflect what are the serious problems, but that it cannot be expected to ensure complete safety from fire – an ideal which is impossible to attain.

7.1.3 Balances and benchmarks

Once all the risks and precautions have been identified and an estimate made of their value, it is then possible to attempt the third part of an assessment. This is consideration of the extent to which the specific risks have been reduced or mitigated. Have the identified risks been balanced by adequate fire precautions? The result of the assessment can be put in two ways.

1. What level of risk remains?
2. What level of precautions exists?

These two ways are of course 'mirror images' of each other and represent the assessment of fire safety of the building.

Once the fire safety of the building has been measured in this way, it needs to be put into context. This can be in one of two forms, either the relative assessment of one building against another, or the absolute assessment of a building against a 'benchmark'. Such a 'benchmark' (or datum) should be an established and recognized standard either in legislation or in guidance. Relative fire assessments can be of great value in the estate management programmes of large organizations. Where a large number of premises are being managed by one body (retail chains, health authorities, government departments) it is important that resources for estate improvement are used to maximum advantage and that the worst premises are targeted as priorities for improvement.

Of course it is preferable to bring all buildings up to a recognized standard, and this can be considered as the second type of assessment, where the building in use is being assessed against such an established standard. With an existing building this will still require assessment to determine if the precautions mitigate or compensate for the risks which have been identified. If the conclusion is that the level of precautions is inadequate to the risk, then the strategies available for improvement need to be considered.

Rarely will there only be one option available to bring a building up to the required standard. Normally there will be a number of alternatives which need to be investigated. Inevitably the building owner will want to achieve the required level with the minimum expenditure, and hopefully with minimum effect on the architectural qualities of the building. Some assessment schemes have been designed to offer alternative strategies for improvement and so assist the designer in choosing the most 'safety effective solution'.

The next four sections outline four specific fire assessment schemes developed by the authors for different building types, and they show the variety of methods and forms of assessment. The reasoning behind the different schemes has been included so that it is possible to see how these have developed from first principles and precisely what 'benchmarks' have been adopted.

7.2 Hospitals

The first assessment method concerns hospitals. Fire safety in hospitals is a very sensitive topic; sensitive both morally and politically as the very existence of hospitals controlled by a national system suggests that the nation, through government, has taken responsibility to care for people who are ill in some way. If patients or staff are to be injured by some external agency such as fire, then this is a direct reflection on the quality of the management of the total health care system.

In the early 1980s the Department of Health and Social Security developed a scheme to assess the relative fire safety of hospital wards. This offered a framework against which the acceptability or otherwise of standards within hospitals could be assessed, but it was not legally enforceable and remained only as a guidance document. However, it was widely used and proved to be useful in identifying hospital wards which posed the greatest risk to life and which therefore most needed improvement. The absence of enforceable standards of fire safety in patient areas became obviously absurd when fire authorities began to enforce standards on to office areas within wards (arguing that these were covered under the Fire Precautions Act), while having to ignore the risk to the much more vulnerable patients.

However, the situation changed in the early 1990s with the need for hospitals to comply with the requirements of European Directives, and the new version of Health Technical Memorandum (HTM) 86 (Fire Risk Assessment in Hospitals) outlines one way of assessing the fire risk of a hospital against the 'benchmark' standards set out in the 'sister' publication (HTM 85: Fire Precautions in Existing Hospitals). These documents are enforced by the fire authorities on behalf of the Home Office under the workplace guidance, as well as being requirements of the Department of Health. Fire engineering principles underlie the new HTMs, and the assessment permits the adoption of alternative fire engineering strategies to achieve an equivalent level of fire safety. In this way the most 'safety effective' solution can be identified.

To evaluate the existing fire risks and fire precautions, the hospital under consideration is divided into a series of assessment areas. These will be determined by the functional layout of the hospital, and normally each nursing and/or other management unit will be an assessment area. The assessment method follows the standard three-part approach already outlined.

7.2.1 Risks

The first part of the assessment is an estimate of the life risk and the hazards in the assessment area using the first 12 worksheets. The first examines life risk, and the rest the fire hazards grouped under the headings 'ignition sources' and 'combustible materials'. The hazards include both design and management issues. They are as follows.

- Life risk
  1. Patients
- Ignition sources
  2. Smoking
  3. Fires started by patients
  4. Arson
  5. Work processes
  6. Fire hazard rooms
  7. Equipment
  8. Non-patient access areas
  9. Lightning
- Combustible materials
  10. Surface finishes
  11. Textiles and furniture
  12. Other materials

These risks and hazards are categorized in the assessment as follows.

- **Acceptable**: the 'benchmark' standard stated in HTM 85.
- **High** or **very high**: as stated on worksheets. High and very high levels of life risk or hazard have to be compensated for by providing higher standards of fire precautions, or they have to be reduced.

7.2.2 Precautions

In the second part of the assessment the remaining 21 worksheets are used to assess the fire precautions in the assessment area. These are grouped under the headings of the five fire safety tactics and include both design and management issues. They are as follows.

- Prevention
  13. Management

14. Training — *staff , student*
15. Fire notices and signs — *fire exit*
- Communications *no smoking*
  16. Observation — *smoke, fire alarm.*
  17. Alarm and detection systems —
- Escape
  18. Single direction escape
  19. Travel distance — *the longest time — for 3rd fr.*
  20. Refuge
  21. Stairways
  22. Height above ground level — *the bldg.*
  23. Escape lighting
  24. Staff — *the staff will guide.*
  25. Escape bed lifts
- Containment
  26. Elements of structure
  27. Compartmentation
  28. Subdivision of roof and ceiling voids
  29. External envelope protection
  30. Smoke control — *sky light*
- Extinguishment
  31. Manual firefighting equipment —
  32. Access and facilities for the fire brigade
  33. Sprinklers

The existing fire precautions are categorized in the assessment as follows.

- **Unacceptable**: precautions which have to be improved.
- **Inadequate**: precautions which are below the 'benchmark' standard stated in HTM 85 (these must either be improved, or compensated for by higher standards in other precautions).
- **HTM 85 standard**: precautions which are provided to the 'benchmark' standard stated in HTM 85.
- **High standard**: precautions which are provided to a higher standard than that stated in HTM 85 (these may be used to compensate for deficiencies in other precautions, or for high or very high levels of risk or hazard).

7.2.3 Balances and benchmarks

Having made the assessment using the worksheets, an assessment record needs to be completed. This indicates the extent to which the existing safety precautions compensate for the risks to life and property identified. Figure 7.2 shows a completed assessment record for a typical hospital ward.

The third part of the assessment process is the identification of the improve-

Assessment record

Worksheets	High standard	Acceptable risk/hazard or HTM 85 standard	High risk/hazard	Very high risk/hazard	Inadequate	Unacceptable
1 Patients				✓		

HAZARDS

Ignition sources   L   M   H

	High standard	Acceptable risk/hazard or HTM 85 standard	High risk/hazard	Very high risk/hazard	Inadequate	Unacceptable
2 Smoking *staff & students*		✓	✓			
3 Fires started by patients		✓	✓			
4 Arson		✓	✓			
5 Work processes		✓				
6 Fire hazard rooms		✓				
7 Equipment		✓				
✓ 8 Non-patient access areas		✓				
9 Lightning		✓				

Combustible materials

	High standard	Acceptable risk/hazard or HTM 85 standard	High risk/hazard	Very high risk/hazard	Inadequate	Unacceptable
10 Surface finishes		✓				
11 Textiles and furniture		✓				
12 Other materials		✓				

PRECAUTIONS

Prevention

	High standard	Acceptable risk/hazard or HTM 85 standard	High risk/hazard	Very high risk/hazard	Inadequate	Unacceptable
13 Management		✓				
14 Training		✓				
15 Fire notices and signs		✓				

Communications

	High standard	Acceptable risk/hazard or HTM 85 standard	High risk/hazard	Very high risk/hazard	Inadequate	Unacceptable
16 Observation					✓	
17 Alarm and detection systems		✓				

Means and escape

	High standard	Acceptable risk/hazard or HTM 85 standard	High risk/hazard	Very high risk/hazard	Inadequate	Unacceptable
18 Single direction escape					✓	
19 Travel distance	✓					
20 Refuge	✓					
21 Stairways	✓					
22 Height above ground level					✓	
23 Escape lighting	✓					
24 Staff		✓				
25 Escape bed lifts	✓					

Containment

	High standard	Acceptable risk/hazard or HTM 85 standard	High risk/hazard	Very high risk/hazard	Inadequate	Unacceptable
26 Elements of structure		✓				
27 Compartmentation	✓					
28 Subdivision of roof and ceiling voids		✓				
29 External envelope protection		✓				
30 Smoke control	✓					

Extinguishment

	High standard	Acceptable risk/hazard or HTM 85 standard	High risk/hazard	Very high risk/hazard	Inadequate	Unacceptable
31 Manual firefighting equipment		✓				
32 Access and facilities for the fire brigade	✓					
33 Sprinklers		✓				

Figure 7.2 HTM 86 – completed record of assessment

ments (if any) which are required to achieve an acceptable level of fire safety. This is done using the compensation sheet, which can be used after the assessment has taken place for two purposes:

1. to establish if higher levels of risk or hazard, or deficient precautions, are compensated for by existing 'high standards' in other precautions;
2. to determine options for improving the fire safety, where higher levels of risk or hazard, or deficient precautions, are not compensated for by existing 'high standard' in other precautions.

Figure 7.3 shows a completed compensation sheet showing the options for improving the ward which is the subject of the assessment record in Figure 7.2.

The left-hand side of the first matrix lists all the risks and hazards which might have been rated as 'high' or 'very high'. The left-hand side of the second matrix lists all the precautions which might have been recorded as 'inadequate'. At the top of both matrices are listed the precautions where a 'high' standard might serve as compensation.

To determine which precautions would compensate for a particular high level of risk or deficiency, look along the row concerned.

• If a box is fully hatched, then a 'high standard' of the precaution in that column will fully compensate for the life risk, fire hazard or deficient fire precaution.
• If a box is partially hatched, then this, together with other partially hatched boxes, collectively represent a group of 'high standards' which will compensate for the life risk, fire hazard or deficient fire precaution.
• If the box is blank, then this precaution cannot be used as a compensating factor.

7.3 Public sector dwellings

Most of the work on the development of assessment methods has concentrated on public buildings and places of work. However, far more people die in fires in their own homes than anywhere else. The statistics in Chapter 1 show clearly that the majority of fires and fire deaths occur in dwellings. Yet the majority of fire safety legislation, guidance and codes of practice is still concerned with non-domestic properties such as offices, shops and hotels. The Welsh Office was particularly concerned at the number of recent fires in Wales with tragic consequences, and which had alerted public attention to the potential dangers; they therefore financed research into the development of such an assessment scheme for public sector housing in Wales. The objective was to produce a simple method of assessing the relative fire risks and safety precautions in existing dwellings. This had to be available as a simple, easy-to-use booklet which housing staff working in the public sector could use as part of their regular inspection process. Through its

Compensation sheet

Precautions which can be used to compensate

Matrix one – Life risk and hazards			16 Observation	17 Alarm and detection	⑲ Travel distance	⑳ Refuge	㉑ Stairways	22 Height above ground	㉓ Escape lighting	24 Staff	㉕ Escape bed lifts	㉗ Compartmentation	㉚ Smoke control	㉜ Fire brigade	33 Sprinklers
①	Patients	– high			●	●			●		●				
1	Patients	– very high													
3	Fires started by patients	– high													
3	Fires started by patients	– very high													
6	Fire hazard rooms	– very high													
11	Textiles and furniture	– high													
11	Textiles and furniture	– very high													

Matrix two – Inadequate precautions		16	17	⑲	⑳	㉑	22	㉓	24	㉕	㉗	㉚	㉜	33
⑯	Observation			●	●									
17	Alarm and detection systems													
⑱	Single direction escape											●		
19	Travel distance													
20	Refuge													
21	Stairways													
㉒	Height above ground level				●	●					●		●	
26	Elements of structure													
29	External envelope protection													

Figure 7.3 HTM 86 – completed compensation sheet

systematic use, housing organizations (local authorities and housing associations) would then be able to compare risks in different housing types and determine priorities for improvement, thus maximizing the effectiveness of the limited resources available.

Local authorities and housing associations are responsible for the standards of safety in their properties and are continually involved in improvement programmes to maintain safety standards. It is reasonably easy to ensure that new buildings achieve an acceptable level of safety, since they must comply with the Building Regulations and the relevant British Standards. However, problems can arise with existing dwellings, as these may fall below the current standards prescribed for new buildings. Given the limited budgets for remedial improvement, housing authorities are inevitably faced with assessing which dwellings pose the greatest risks and are therefore the priorities for improvement. The assessment method has been designed to enable housing authorities to take concentrated remedial action where the risk is greatest, and therefore hopefully to reduce the incidence of domestic fires.

Although designed for use in public sector dwellings within Wales, the basic ideas behind the scheme are applicable to most dwellings. It is now hoped to attract funding to develop versions for other parts of the British Isles, and for the private sector dwellings. The current method is also restricted to low-rise single dwelling units. This was because of the special problems inherent in high-rise developments, sheltered accommodation and special needs housing. However, it is intended to consider these particular housing types in future research.

The research used a combination of published information and the experience of an expert advisory group. There was neither the time nor the funds to embark on a vigorous inquiry into the causes of all fires occurring in public sector housing in Wales, nor were there sufficient data available from past dwelling fires to develop an assessment of risks based on a simple statistical exercise.

In a series of meetings the group identified risks and precautions related to domestic fire safety and drafted 'good practice standards' for each of these factors to act as a 'benchmark'. Obviously some risks and precautions were related to the tenants' life-styles, while others were clearly the responsibility of the housing managers. Both the housing management and the tenants have a role in achieving fire safety, and it was quickly recognized that the assessment scheme would have to be in two parts.

The first part looks at the dwelling itself and at the risks inherent in the home, concentrating on the design and structure of the property (the design issues). The role of the housing organization in promoting fire safety is also included at the end of Part 1 (the management issues). By using Part 1, housing staff should be able to make an assessment of the relative safety of different dwellings. Twenty factors are assessed; these and their relative values are as follows:

1. Tenants   6%
2. Heating   6%

3.	Electricity and gas	6%
4.	Smoke alarms	6%
5.	Outside walls	6%
6.	Escape routes	6%
7.	Electrical sockets	5%
8.	Windows	5%
9.	Height above ground	5%
10.	Fire brigade access	5%
11.	Fire protection between homes	5%
12.	Lofts, ducts and cavities	5%
13.	Number of homes sharing escape routes	5%
14.	Fire resistance of shared escape routes	4%
15.	Storage	4%
16.	Inside walls and ceilings	4%
17.	Public telephones	3%
18.	Information for tenants	5%
19.	Fire safety training for housing staff	4%
20.	Maintenance and fire safety	5%

Figure 7.4 shows a completed Part 1 assessment sheet for a dwelling.

The second part is for tenants to complete by themselves. It is designed as a checklist which highlights the most common risks and precautions in their own homes, and includes both 'design' and 'management' issues. It consists of a series of questions which need to be answered 'Yes' or 'No'. Where the response is 'No' then this signals to the tenants that there might be a problem, and they are encouraged to minimize the number of 'No's. The factors assessed in Part 2, the tenants' fire safety checklist, are as follows (they are not given relative values):

21. Electrical appliances
22. Chip pans and cooking
23. Smoking
24. Open fires
25. Portable heaters
26. Foam filled furniture
27. Wall papers, tiles and paints
28. Dangerous materials
29. Blocking escape routes
30. Fire plans
31. Smoke alarms
32. Fire blankets

The final assessment scheme is in the form of a 12-page booklet, written as simply as possible, and offering as much help as it can to the assessor. It must

---

Score sheet for Part 1: The Home Assessment

During the assessment process this sheet can be used in addition to the booklet to record the comparative safety of the home. For each question in the booklet read the text carefully and tick the most appropriate box. Only one box may be ticked per question. Add the scores together to make a comparative assessment of the level of fire safety and record this final figure in total box.

1	Tenants	0	[4]	6
2	Heating*	0	3	[6]
3	Electricity and gas*	[0]		6
4	Smoke alarms	[0]		6
5	Outside walls*	[0]		6
6	Escape routes	0		[6]
7	Electrical sockets	0		[5]
8	Windows	[0]		5
9	Height above ground	[0]	2	5
10	Fire brigade access	0		[5]
11	Fire protection between homes*	[0]		5
12	Lofts, ducts and cavities*	0		[5]
13	Number of homes sharing escape routes*	[0]	3	5
14	Fire resistance of shared escape routes*	[0]		4
15	Storage	0		[4]
16	Inside walls and ceilings*	[0]		4
17	Public telephones	[0]		3
18	Information for tenants	[0]		5
19	Fire safety training for housing staff	0		[4]
20	Maintenance and fire safety	[0]		5

Total score **39**

* Issues where technical knowledge may be necessary. If in doubt seek guidance.

Figure 7.4 Public sector dwellings – completed Part 1 record sheet

be remembered that the scheme was not intended to be used as a precise, accurate tool. It was just a relatively crude method of relative fire assessment for single unit dwellings within the public sector in Wales. If used properly it can assist housing organizations in maintaining safety standards in existing properties, and can alert tenants to possible fire issues within their own life-style.

7.4 Workplaces

As has been outlined at the start of this chapter, the need to implement the requirements of two key European Directives acted as a spur to the development of fire assessment methods. For architects these directives will introduce a number of profound changes, as they extend the remit of the existing fire safety legislation to cover not just the design issues (e.g. the fabric and layout of a building), but also management issues (i.e. the way it is used and managed). They also significantly extend the range of buildings covered to include many previously outside the Fire Precautions Act (such as hospitals, schools, churches).

Responsibility for undertaking such assessments is given in most cases to the owners, and there is great emphasis on self-regulation and compliance. However, in new buildings or improvement projects architects might well find themselves being required to undertake such assessments or, what is more likely, being asked to confirm to their clients that such assessments have already been undertaken as part of the design process.

The Home Office have yet to produce guidance on the nature of such fire assessments. This is a new field and architects would be well advised to be aware of potential dangers and difficulties in such appraisals.

In 1993 the authors produced for the Home Office a proposed structure for fire assessment and this is outlined in Figure 7.5. This assessment scheme was kept simple in order to make it accessible to the ordinary employer, and as assessment is dependent on the activity in the workplace, the built form, the staffing and the management, it was stressed that a change in any of these would require a new assessment. The assessment had five simple stages which looked at hazards, the possibilities for reduction, the mitigating influence of the fire precautions, the risks to people in the workplace, and finally determined the risk category.

The first stage of the assessment analysed the fire hazards using two work-sheets. One considered materials which burn easily and were present either as contents or in the fabric of the building, and the other looked at potential sources of ignition. The hazards were classified as either acceptable or unacceptable.

The second stage involved considering possible reductions of the hazards now identified. This was often by simple and inexpensive methods and if successful would lead to the risk category of the workplace being adjusted. In the third stage the existing precaution measures were evaluated to determine if they would compensate for the remaining fire hazards. Additional safety measures could be

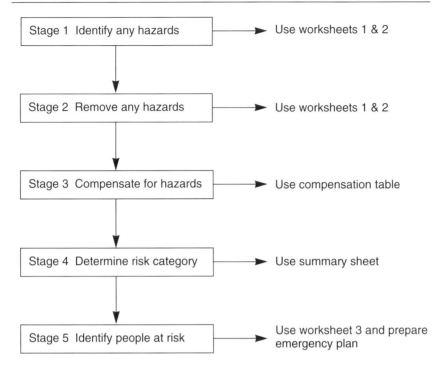

Figure 7.5 Fire assessment of the workplace: methodology developed for the Home Office, 1993.

identified at this stage, although their installation might need to be the subject of a cost–benefit appraisal.

Stage four involved the determination of the final risk category. Having identified the hazards, taken into account any possible reductions, and the value of existing or additional measures in compensating for the hazards, the employer should be able to determine the risk category of the workplace.

The last stage of the assessment was the analysis of the risk to people using the workplace through their responsiveness, mobility, awareness and familiarity. This was done using the third worksheet, and on the basis of this the emergency plan could be produced.

7.5 Canals

The final example refers not to a building at all, but to an unusual structure which possibly provides one of the most potentially dangerous environments imaginable – the canal tunnel. British Waterways are responsible for the maintenance of almost all canals in Great Britain, and although commercial use of the canal system is now restricted to limited areas, there has been a rapid increase in the use of the canals for recreation and holidays. On the waterway network there are over 40 major tunnels which are still navigable, and the longest of these is nearly 3000 m. All are very narrow, and most have no towpaths or walkways; they are generally unlit and unventilated, and may well have traffic passing through them in both directions. Many of the boats using the tunnels are hired for a single week by inexperienced and untrained people, and in some tunnels there are also 'trip' boats or restaurant boats which may carry up to 80 passengers. The tunnels were built between 1770 and 1858, and although structural inspections are regularly conducted, there have until now been no fire or other safety assessments.

The safety record in canal tunnels has been fairly good, and there certainly have not been any major disasters. However, in 1990 British Waterways decided to investigate the safety of people using these tunnels before any disaster happened. As part of this research programme they decided to develop a minimum safety standard (effectively a 'benchmark') for canal tunnels and an assessment method to evaluate individual tunnels against this standard and determine priorities for improvement.

Assessment is made of ten precautions; these and their relative values are as follows.

1.	Access/egress within the tunnel	26%
2.	Communications	20%
3.	Pre-planning	10%
4.	Access to tunnel portal	9%
5.	Signs and notices	7%
6.	Traffic management	7%
7.	Monitoring surroundings	4%
8.	Lighting	8%
9.	Ventilation	7%
10.	Anti-vandal	2%

As it is a relative assessment of the tunnels, the risks could be considered to be constant, and therefore they did not need to be included in the assessment method.

The minimum safety standard covers all the ten precautions as well as five further issues which were considered to be general to the whole waterways system, rather than varying between different tunnels. As these were a constant they were excluded from the relative assessment of the individual tunnels. These non-specific precautions were:

- users' education
- monitoring
- boat standards
- boat contents
- user regulations

The minimum safety standards and assessment methods together provided a simple, coherent and co-ordinated strategy for assessing and improving canal tunnel safety. Figure 7.6 shows a completed assessment sheet for a canal tunnel. As a result of the research it was possible to recommend that a programme of upgrading be instituted in order to bring all tunnels up to the minimum safety standards, the order of priority for such upgrading being determined by using the assessment method to calculate the 'collective risk exposure score' for each tunnel. This takes into account the presence or absence of precautions, the length of exposure (how long it takes to pass through the tunnel) and the number of people using the tunnel each year.

The assessment method has become part of an annual safety audit conducted on each tunnel and its surroundings, and linked to the series of regular structural

---

Summary sheet

Name of tunnel: **Tardebigge**

Length (km): **0.53**

Usage (1000s pa) **45**

Surveyors:

Date:

Practical safety measure	Relative value		Score		Total
1 Grab rails	4	×	⟨-2⟩ -1 0 1 2 3	=	−8
2 Refuges	5	×	⟨0⟩ 1 2 3	=	0
3 Towpaths	8	×	⟨-2⟩ -1 0 1 2 3	=	−16
4 Rescue boats	8	×	⟨-2⟩ -1 0 1 2 3	=	−16
5 Ventilation	7	×	⟨0⟩ 1 2 3	=	0
6 Lighting	9	×	⟨0⟩ 1 2 3	=	0
7 Communication (boat/rescuers)	10	×	-2 -1 ⟨0⟩ 1 2 3	=	0
8 Communication (rescuers)	10	×	-2 -1 ⟨0⟩ 1 2 3	=	0
9 Portal access	9	×	-2 ⟨-1⟩ 0 1 2 3	=	−9
10 Signs and notices	7	×	-2 -1 ⟨0⟩ 1 2 3	=	0
11 Anti-vandal	2	×	⟨0⟩ 1 2 3	=	0
12 Pre-planning	10	×	-2 -1 ⟨0⟩ 1 2 3	=	0
13 Traffic management	7	×	⟨-2⟩ -1 0 1 2 3	=	−14
14 Monitoring surroundings	4	×	-2 -1 0 1 2 ⟨3⟩	=	12

Safety score   100   (−154 to 300)   =   −51

Risk score: −63

Risk exposure score: −33.39

Collective risk exposure score: −1502.55

NOTES:

Figure 7.6 Canals – completed assessment sheet

inspections already in place. Collective risk exposure scores for the tunnels will vary with time, and it is essential that the individual tunnels are kept under review.

7.6 Audits

Audits should not be confused with the more familiar assessments, appraisals or evaluations. While assessments estimate the fire risks and fire precautions within a building, the purpose of an audit is much more limited. It is an examination of the assessment (which must already exist) to check its accuracy in the estimation of fire risks, and the appropriateness and implementation of the fire precautions. As such it will involve a review of the management procedures to ensure that the existing fire safety policies are both correct and are being implemented. Full assessments normally involve a structural survey, and are therefore too expensive and time consuming to be undertaken at very frequent intervals. The audit procedure is a much simpler desktop exercise which can be undertaken regularly, perhaps once a year, and whenever there is a change in building use or an unusual event.

Before any type of audit it is necessary to be clear about the management structure which is being audited. Three distinct groups should be identified: the policy makers, the advisers and the operational staff. In the case of a fire audit of a cathedral, which will be considered in the next subsection, it is normally the Dean and Chapter who are the policy makers, for it must be they who carry the ultimate responsibility for deciding priorities and policies. In the case of hospitals, which are also considered later, then it is clearly stated that the policy makers are the management board and the Chief Executive, or General Manager. Such policies must cover the pro-active fire safety measures which should always be in force, the fire emergency plan itself and the post-fire recovery strategy.

Experts can provide advice and should draft such policies, but they can only be implemented with the full knowledge and consent of the policy makers. In a cathedral the advisers are likely to be the cathedral architect or surveyor to the fabric, the insurer, the fire brigade and, hopefully, a specialist fire engineer. In the case of a hospital, there is a formal job description for the 'specialist fire safety adviser' provided in the national guidance.

The third group are those who have to implement the policies once made, and these are the people who are responsible for the day-to-day running of the organization. They are the ones who must see that the pro-active measures are maintained, and will initially have to take charge in the event of a fire. In the case of cathedrals with their own works staff, this role can be ably fulfilled by the clerk of works and the site staff and it may be volunteer guides or helpers who find themselves in this role. In hospitals this role is expected to be filled by the nominated officer (fire), and that person's duties both in normal conditions and in fire situations are again defined in national guidance.

7.6.1 Fire safety audits of cathedrals

One can take the fatalist view that fires in major churches are so rare and so hard to prevent, that it is not worth investing large resources in fire prevention. However, in the last few years there have been cathedral fires in Nantes (1972), York Minster (1984), Perugia (1985), Luxemburg (1985) and Newcastle (Roman Catholic) (1988). The potential for loss is huge, and therefore the small investment involved in regular auditing is easily justified as a way to maximize the opportunities for fire safety.

In preparing a fire safety audit of a cathedral, each of the three main policy areas should be reviewed and evaluated. The first of these comprises the pro-active measures which have been put in place to reduce or compensate for identifiable risks. The audit will involve checking that all hazards have been identified and appropriate mitigation measures taken. Hazards come in the form of potential ignition sources (e.g. lightning, electrical or mechanical faults, human error, deliberate fire raising) and combustible materials (e.g. storage, bookshops, sacristy, seating, rubbish, building fabric). The mitigation measures will include both design issues (e.g. lightning protection, electrical safety systems) and management issues (e.g. control of access to different parts of the building, working practices). In an audit the existence and appropriateness of both these types of measures are checked. Maintenance and testing can be reviewed, staff awareness of procedures tested and the suitability of measures to specific events evaluated.

As well as the risks the pro-active policies should cover the precautions which should be in place to limit the danger to people and to the building should a fire start. The first stage is the alerting of the occupants and fire brigade, and the audit will review the chosen systems to check they are still appropriate and that they are being maintained and tested correctly. The planning and, if necessary, sign-posting of the means of escape also has to be checked. A change in the way services or events are held may require a revision to the planned escape routes and the training of the staff in alternative evacuation procedures. The audit will also review any recent activities in the building (e.g. general maintenance, minor alterations) to ensure that none of the designed containment measures has been invalidated or breached. Obviously an audit would also involve a check on the maintenance and testing of the manual firefighting equipment, and of any fixed systems which it has been possible to install.

The second policy area which the audit reviews is the fire emergency plan itself; this is what is planned will happen when a fire does occur. There must be a series of such plans to take into account the most likely fire scenarios, and these should have been worked out in conjunction with the fire brigade. All staff (paid and volunteer) need to be aware of what part they will have to play, and the whole system should have been tested as far as is practical in exercise. The emergency plan will cover alerting the fire brigade, evacuating the building (and this might be difficult from areas such as crypts or towers) and any special plans to protect particular works of art. The fire emergency plan must also cover the essential immediate post-fire activities (e.g. reducing water loading on vaults, recording rather than clearing of debris, specialized salvage).

The final set of management policies which should be regularly audited is the strategy for repair and replacement after a fire. This strategy has to be reviewed to ensure that the assembly and storage of archival materials is sufficiently comprehensive to permit replacement (e.g. drawings, photographs, ornament samples, plaster casts, photogrammetric surveys, archives). These will of course need to be stored away from the cathedral itself! Thought should also have been given in advance to what will be the priorities for work, what equipment may be needed, and where specialist staff can be obtained. An inventory of works of art (embroideries, pictures, metalwork, manuscripts, etc.) will already be in existence for the insurers, but this needs to be reviewed so that it includes contemporary conservation and restoration information about each item.

7.6.2 Fire safety audits of hospitals

While in cathedrals the fear is property loss, in hospitals the priorities are different, and it is life safety which must always be the paramount concern. The issues of fire safety assessments have been considered earlier in the chapter, and the requirements for annual audits make clear the difference between these two activities.

> The processes of a fire safety audit differ from those for a risk assessment
> . . . Risk assessment firstly identifies fire hazards and the particular risks they present to the occupants of a premises. Following a thorough assessment of the risks, effective fire precautions are arranged to match the level of fire risk. A fire safety audit would verify that these fire precautions, once in place, are being maintained effectively. (Health Technical Memorandum (HTM) 83)

The requirements for audits are specified in the national guidance. They

> should examine and question all aspects of fire safety. They may be carried out either by competent staff employed by the health authority/trust or by external consultants. The audit should be systematic and cover all aspects of fire safety, including physical precautions, staffing arrangements and management systems. Where required, validation checks must be included in the audits. For example documentary evidence supporting the fire precautions policy should be examined, the visual integrity of cavity barriers, fire stopping etc. should be verified, and remedial actions set in train where necessary. The fire audit team must have full access to the relevant staff, records, buildings and plant. (HTM 83)

7.7 Assessors

Having examined some of the principles which must underlie fire assessment and considered a number of specific examples this final section examines the problem of finding assessors who are competent to carry out these assessments. It will not be enough just to have assessment schemes and guidance documents, there must also be assessors capable of using these materials.

In the case of the hospital assessment documents and for hospital fire audits it is stated that

> (these) documents may be used by competent staff ... staff will be considered competent where they have sufficient technical training and actual experience or technical and other qualities, both to understand fully the dangers involved and to undertake properly the measures referred to ... (HTM 86)

This definition of competence follows closely that included in the Management of Health and Safety at Work Regulations, where it is stated that

> a person shall be considered competent ... where he has sufficient training and experience or knowledge and other qualities to enable him properly to assist in undertaking the measures ...

The scheme for the assessment of public sector dwellings was designed with the awareness that it would be used by tenants, and the second part was deliberately planned as a simple checklist. Although no experience could be assumed when doing individual assessments, it was stressed that if the findings from an authority's entire housing stock were to be used as the basis for improvement programmes then it was essential that the staff employed by the housing management organizations be competent to deal with the complexities of relative assessment. The third example of the scheme designed by the Home Office for use in the workplace has also been designed to be used by the non-technical employer. However, again it is stressed in the documents that in complex buildings, or where the work processes are in any way hazardous, it will be necessary to seek expert advice. In the final example it was assumed from the very start that the canal tunnel assessment method would only be used by experienced British Waterways employees and that therefore a fair amount of knowledge and experience could be assumed.

The extension of fire safety engineering into the field of fire assessment raises a series of fundamental questions about competence to undertake such assessments. Assessment has to involve a recognition of varying levels of risk and precaution. It cannot be a simple series of prescriptive standards for application to all buildings irrespective of what is happening inside them. It must be more sensitive to variations in risk than the crude categorization of buildings into purpose groups on which the Approved Document to the Building Regulations depends.

While almost everyone now acknowledges the importance of fire engineering, who are the fire engineers competent to do these 'safety effective' assessments and ready to provide impartial professional advice to the ordinary building owner about fire safety? Until recently there has been no clear answer to this problem, for in the case of the United Kingdom three distinct types of fire engineers had developed, each fulfilling a useful function that was often carried out in conflict or mutual suspicion rather than collaboration.

The first group of fire engineers were those employed by the statutory authorities, those who enforce the Building Regulations and Fire Precautions Act. Most of these people are experts on the legislation and can often quote precise chapter and verse from the myriad government guidance and advisory documents to support their position. Many of them in the fire brigades have long years of experience of fighting fires and know the terrible consequences of inadequate fire precautions. Those in the building control departments have excellent experience of the realities of the construction trade and the real difficulties of ensuring that contractors do build to all the specified standards. These fire engineers probably belong to the Institution of Fire Engineers and are rightly proud of having passed their respective exams.

The second group of fire engineers were those involved in the business of designing and installing the active precaution systems which can add to the safety of a building: the alarm and detection systems, the sprinkler systems and the smoke extract and ventilating systems. They probably had a mechanical or electrical engineering background, and until recently would probably have been members of the Society of Fire Prevention Engineers. These fire engineers often play an invaluable role on the various British Standards Institution committees attempting to set standards for all this hardware.

The third group of fire engineers, perhaps the fewest in number, were the ones most involved in the scientific analysis of fire behaviour and fire science. Often they have worked for the Fire Research Station or one of the growing number of universities involved in fire safety. They have a concern for systems and models of fire safety and encompass all the scientific disciplines from human behaviour to applied chemistry. Many will have research degrees, and most would have, until recently, belonged to the third large fire association, the Society of Fire Safety Engineers.

These three groups have over the last few years moved much closer together, and the merger process was completed in 1998. The Institution of Fire Engineers is now the sole body in the United Kingdom representing the profession. Most importantly this body now has an Engineering Council Division, which has the approval of the Engineering Council and therefore gives its members the coveted status of chartered engineer. Most of the former members of the Society of Fire Prevention Engineers and the Society of Fire Safety Engineers are now members of this Engineering Council Division.

Fire assessments demand competent individuals who can combine the skills of all three earlier bodies. It will be necessary to know not only the contents of the legislation and guidance, but also the basis on which these documents were developed. This will then have to be combined with a knowledge of possible hardware systems, and an understanding of the underlying principles of fire assessments. With the establishment of a group of chartered fire engineers, architects and designers now have a simple method of establishing just who is competent.

Information

8

8.1 Introduction

This chapter is intended to guide both architects and the staff of statutory authorities to the information they may need in working from first principles. There is so much information available that there is a real danger that important issues can become lost in a plethora of insignificant and confusing detail. This chapter cannot summarize all this information, instead it should serve as a guide to that which is most important for those working within the United Kingdom.

The next two sections consider the legislation relevant to the different parts of the United Kingdom: Great Britain and Northern Ireland. The relationship between Acts and regulations/codes is examined and the application and authority of the various documents is outlined. The precise details of the legislation are not reproduced as designers must have their own copies of the Approved Documents, Technical Standards or Technical Booklets, but an attempt is made to explain the relevance and application of the different parts of the legislation.

Much of the legislation refers directly to the British Standards, and these are introduced in section 8.4. These documents occupy a curious position, almost akin to legislation in their authority and it is essential for the designer to be aware of the most important standards and their different numbers and codes.

Section 8.5 summarizes all the other guidance available from research, government and trade bodies. This includes a review of other textbooks which are available. A list of sources of advice is also provided in section 8.6.

The penultimate section consists of a glossary of fire safety terms used in this book and commonly used in building design. The last section is an index to the whole book.

8.2 Legislation in Great Britain

Most of the legislation relating to fire safety within Great Britain has been enacted as a result of tragedies or particular fires. It has therefore developed in a piecemeal fashion with different sets of legislation referring to different building types and distinguishing between new buildings and existing buildings. The process of refinement and replacement has given the law an almost geological quality with act and regulation superseding the previous act and regulation sometimes completely, as in a consolidating Act, sometimes only partially. This incremental process has left some gaps and created some areas of overlap. It can also be very confusing and is not always consistently logical. This is another reason why the designer must consider fire safety from principles and not from legislative compliance. Although the development of the legislation is probably only of passing interest to the designer, various books give a good historical perspective of fire safety law, notably Read and Morris, Third Edition, 1993, *Aspects of Fire Precautions in Buildings.*

One of the more recent influences on the formation of fire safety legislation has been the need to comply with European directives, and the Fire Precautions (Places of Work) Regulations are perhaps the first of a series of such initiatives.

Not all documents have the same status and it is important for the architect to be aware that many of those to which statutory authorities may demand slavish obedience, are only advisory. Legal requirements can be categorized in order of their significance as:

1   Acts of Parliament (or Enactments) – these are the laws passed by Parliament which have to be obeyed. However, these very rarely contain technical details, and normally only set up procedures or empower government departments to make regulations.
2   Regulations (or Statutory Instruments) – made by government departments under enabling acts, these have the force of law. They used to contain technical information, but in recent years the actual regulations have been given instead as functional requirements. In Scotland the prescriptive Technical Standards which are the only method of fulfilling the requirements of the regulations also have this status.
3   Approved Documents and Guides – these do not constitute the law. They are normally simply advisory documents which suggest one method of fulfilling the requirements of the regulations.

British legislation can be divided into two broad categories: that dealing with new buildings and that dealing with existing buildings. New buildings are primarily the responsibility of the building control departments of local authorities, while existing buildings are primarily the responsibility of the fire prevention departments of fire authorities. The first two of the following subsections deal respectively with new buildings in England and Wales and in Scotland (separate systems having been retained in the different countries). The remainder of the following subsections consider new and existing buildings. The third subsection deals with those buildings to which the Fire Precautions Act applies. The fourth subsection is concerned with the Fire Precautions (Places of Work) Regulations, which apply to almost all workplaces. Following this there are subsections covering the Health and Safety Act, the Licensing Acts and finally information on other legislation which applies to specific local areas or building types.

One of the problems for the architect is the sheer mass and complexity of legislation which could apply; this section does not attempt to summarize the legislation but only to describe the 'patchwork quilt' of overlapping documentation so that the design team can see how the different pieces interrelate.

8.2.1 Building Regulations – England and Wales

The Building Act 1984 consolidated the law relating to building control for England and Wales, but it contains no design information as such. The Act empowers the Secretary of State to make Building Regulations and under it the successive regulations have been issued. The Act also gives local authorities some detailed powers to deal with such matters as adequate drainage, building on filled ground and demolition. In addition, the Act gives local authorities the power (sections 24, 71 and 72) to demand the provision of means of escape in case of certain buildings after consultation with the fire authority. This provision is designed to cover non-domestic buildings where the Building Regulations might not apply.

The current Building Regulations for England and Wales were made under the Building Act of 1984, and Part B covering Fire Safety came into force in its present form on 1 June 1992. There are five functional requirements which apply to all new buildings (except prisons).

B1: Means of Escape
The building shall be designed and constructed so that there are means of escape in the case of fire from the building to a place of safety outside the building capable of being safely and effectively used at all material times.

B2: Internal Fire Spread (Linings)
(1) In order to inhibit the spread of fire within the building, the internal linings shall:
(a) resist the spread of flame over their surfaces; and
(b) have, if ignited, a rate of heat release which is reasonable in the circumstances.
(2) In this paragraph 'internal linings' mean the materials lining any partition, wall ceiling or other internal structure.

B3: Internal Fire Spread (Structure)
(1) The building shall be so constructed that, in the event of fire, its stability will be maintained for a reasonable period.
(2) A wall common to two or more buildings shall be designed and constructed so that it resists the spread of fire between those buildings. For the purposes of this sub-paragraph a house in a terrace and a semi-detached house are each to be treated as a separate building.
(3) To inhibit the spread of fire within the building, it shall be sub-divided with fire resisting construction to an extent appropriate to the size and intended use of the building.

(4) The building shall be designed and constructed so that the unseen spread of fire and smoke within concealed spaces in its structure and fabric is inhibited.

B4: External Fire Spread
(1) The external walls of the buildings shall resist the spread of fire over the walls and from one building to another, having regard to the height, use and position of the building.
(2) The roof of the building shall resist the spread of fire over the roof and from one building to another, having regard to the use and position of the building.

B5: Access and Facilities for the Fire Service
(1) The building shall be designed and constructed so as to provide facilities to assist fire fighters in the protection of life.
(2) Provision shall be made within the site of the building to enable fire appliances to gain access to the building.

Each of the five functional requirements states that the building should have a level of safety, but this is either undefined or only described as 'reasonable'. The Building Regulations are supported by an Approved Document which defines what is the acceptable, or 'reasonable', level of safety through precise specification. This is: Approved Document B, Fire Safety, *The Building Regulations 1991*, Second Edition with 1992 Amendments, Department of the Environment/ Welsh Office, HMSO.

There is no obligation on the designers to follow the solutions outlined in the Approved Document and they are entitled to meet the relevant requirement in any other way which they can demonstrate offers an equivalent level of safety; however, virtually all architects and statutory authorities tend to treat the Approved Document as if it was gospel and treat alternative fire safety solutions with great scepticism. The position of Approved Documents can be compared with that of the Highway Code; it is not an offence not to follow it, but should anything go wrong this will be the standard which the courts would look to as representing reasonable practice. In reality few architects have been brave enough to venture out on their own and the Approved Document has become almost a prescriptive piece of legislation which is obediently followed in virtually all designs.

As the Building Regulations are now in the form of functional requirements it is impossible to apply for relaxations. However, if a local authority and a designer cannot agree on whether or not a proposal fulfils the functional requirement it is possible to apply to the Department of the Environment, Transport and the Regions (DETR) for a determination. This procedure is rarely used because of the time needed to consider the proposal.

The Building Regulations are administered by the local district councils. The architect can either go to them for approval or, in theory, go to an Approved Inspector employed by the architect. Approved Inspectors are increasing in number and work either individually or as corporate bodies. They are beginning to provide real competition to local authorities and with the removal of the scale of fixed fees a free market is beginning to develop.

The Approved Document is structured under the five functional requirements with sections covering the major areas of importance:

B1: Means of Escape
    Section 1   Dwellinghouse
    Section 2   Flats and maisonettes
    Section 3   Design for horizontal escape – buildings other than dwellings
    Section 4   Design for vertical escape – buildings other than dwellings
    Section 5   General provisions common to buildings other than dwellings
B2: Internal Fire Spread (Linings)
    Section 6   Wall and ceiling linings
B3: Internal Fire Spread (Structure)
    Section 7   Loadbearing elements of structure
    Section 8   Compartmentation
    Section 9   Concealed spaces (cavities)
    Section 10 Protection of openings and fire stopping
    Section 11 Special provisions for car parks and shopping complexes
B4: External Fire Spread
    Section 12 Construction of external walls
    Section 13 Space separation
    Section 14 Roof coverings
B5: Access and Facilities for the Fire Service
    Section 15 Fire mains
    Section 16 Vehicle access
    Section 17 Personnel access
    Section 18 Venting of heat and smoke from basements

In addition to Approved Document B, the Basement Development Group led by the National House Building Council produced, in 1997, an Approved Document, Basements for Dwellings, bringing together all issues related to basements, including those covered by Part B.

During 1998 the DETR undertook a consultation process with a view to making some changes to the Approved Document. Therefore a new version of Approved Document B is expected in 1999.

Until the advent of the 1985 Building Regulations the 11 Inner London Boroughs and the City of London retained their own legislative requirements which traced their descent from laws first enacted in the twelfth century. The new Building Regulations were applied to Inner London as from 6 January 1986 and are now administered by the individual London Boroughs. A number of specific regulations were, however, retained and still apply only to Inner London, probably most significant of these being the old sections 20 and 21 which set additional fire safety standards for tall buildings.

8.2.2 Building Standards (Scotland) Regulations

Scotland developed national Building Regulations before England and Wales, and it has remained a distinct and separate system. The Regulations currently in force were made under the Building (Scotland) Act 1959 in 1990 and are functional requirements. They are supported by Technical Standards and Deemed to Satisfy Provisions. The two Regulations relating to fire safety are as follows.

12. Structural Fire Precautions.
Every building shall be so constructed that for a reasonable period, in the event of a fire:
(a) its stability is maintained;
(b) the spread of fire and smoke within the building is inhibited;
(c) the spread of fire to and from other buildings is inhibited.

13. Means of Escape from Fire and Facilities for Fire Fighting.
Every building shall be provided with:
(a) adequate means of escape in the event of fire; and
(b) adequate fire fighting facilities.
Every dwelling shall be provided with means of warning the occupants of an outbreak of fire.

These two regulations can only be complied with by following the Technical Standards. Part D of the Technical Standards relates to Regulation 12 and Part E relates to Regulation 13. As the Technical Standards are mandatory there is not the same potential for alternative approaches that exists in England and Wales. An architect wishing to use a fire safety engineering approach rather than follow precisely the Technical Standard will have to approach the local authority concerned for a relaxation of the standards concerned. Accompanying the Technical Standards are a series of Deemed to Satisfy Provisions. These are not mandatory, but indicate ways of satisfying the Technical Standards which will always comply.

The most recent revision of Parts D and E came into force in December 1997 and was published as the Fourth Amendment to the Technical Standards. Harmonization with England and Wales is a two way process and it is anticipated that the next revision of Approved Document B will further harmonization. However complete uniformity is unlikely without new primary legislation in Scotland, which would permit the Technical Standards to be written in such a way that greater innovation was permitted.

The local authorities administer the building control system in Scotland and there are no Approved Inspectors.

8.2.3 Fire Precautions Act

Unlike the various sets of Building Regulations the Fire Precautions Act of 1971 applies to both new and existing buildings. This Act permits the Home Secretary to designate building types where Fire Certificates must be obtained from the local fire authority, and since the introduction of the Act there have been three designation orders made. The first, in 1972, covered hotels and boarding houses; the other two were in 1977 and covered first factories and secondly offices, shops and railway premises.

For the designated building types the Home Office and the Scottish Home and Health Department have published guidance on the standards that are required for certification. The guidance for hotels and boarding houses is: *Guide to Fire Precautions in Premises used as Hotels and Boarding Houses which require a Fire Certificate* (HMSO, 1991). This guide covers means of escape, walls and ceiling finishes on escape routes, fittings and furniture on escape routes, fire warning equipment, and fire fighting equipment. There is also a shorter guide to management requirements: *Fire Safety Management in Hotels and Boarding Houses* (HMSO, 1991).

The guide for factories, offices, shops and railway premises is: *Guide to Fire Precautions in existing Places of Work that require a Fire Certificate* (HMSO, 1989). This guide sets out the basic standards for means of escape and other related fire precautions in existing premises which require a fire certificate.

There is also a suggested code of practice for buildings too small to require certificates: *Code of Practice for Fire Precautions in Factories, Offices, Shops and Railway Premises not required to have a Fire Certificate* (HMSO, 1989). This gives practical guidance on standards for means of escape in the event of fire and means of fire fighting in existing premises where the requirement for a fire certificate is waived, or there are insufficient numbers for a fire certificate to be necessary.

The Home Office have produced a simple occupiers' and owners' guide to the requirements of the Fire Precautions Act entitled *Fire Safety at Work* (HMSO, 1991). This guidance is intended for owners, occupiers and managers of work places and contains good general fire precautions advice for building designers.

For sports grounds there is the *Guide to Safety at Sports Grounds* (HMSO, 1990).

The fire authority must consult with the relevant local authority before requiring work to be done prior to issuing of a fire certificate. There is also a statutory bar on the fire authority which prevents them from making requirements where at the time of the erection of the building it complied with Building Regulations which imposed requirements as to the means of escape.

Fire certificates specify:

1. Means of escape to be provided
2. Fire fighting and fire escape requirements to be provided
3. Fire alarm system to be provided
4. Limitations on any particular explosive or flammable materials which may be stored on site

They may also cover:

5. Maintenance conditions
6. Staff training
7. Limits on the number of people permitted in the building
8. General fire precautions.

The Fire Safety and Safety at Places of Sport Act of 1987 amended the Fire Precautions Act and established a class of premises which can be exempted from the necessity of obtaining a certificate. However it requires the building's owners/managers to provide an adequate means of escape and firefighting equipment.

Obviously there is a substantial degree of overlap between the role of the fire authority and that of the building control authority especially in regard to new buildings. The different government departments (Environment, Welsh Office, Home Office) have produced a procedural guidance document for designers to clarify the division of responsibility between the statutory authorities and the approvals process which the architect should follow. This extremely useful document applies to England and Wales: *Building Regulations and Fire Safety, Procedural Guidance* (Department of the Environment/Home Office/Welsh Office, 1992). This document is currently being reviewed and a revised version taking account of legislative changes is expected.

The government launched a consultation on the future of fire safety legislation at the end of 1997 and it is likely that this will eventually lead to a new fire safety bill. This will replace the existing Fire Precautions Act, the Fire Precautions (Places of Work) Regulations (see below) and consolidate other fire safety requirements. The intention is to place the responsibility for the assessment of fire risk on the owner or employer and to enable fire authorities to concentrate their resources on the more serious fire risks.

8.2.4 Fire Precautions (Places of Work) Regulations

The Fire Precautions (Places of Work) Regulations were made under the Fire Precautions Act in 1997, however they need to be considered quite separately as their origin and organization is quite distinct from the designation and certification procedures outlined above. These regulations were made in response to the European Council Directives 89/391/EEC and 89/654/EEC known respectively as the framework and workplace directives. They apply to most buildings where people work, but are not intended to place an additional burden on premises already covered by legislation. Therefore the following premises are exempted:

– those which have a fire certificate in force under the Fire Precautions Act,
– those which have a public entertainment licence, and
– those which have a safety certificate, or a special safety certificate, under the Fire Safety and Safety at Places of Sport Act.

Therefore, although the regulations extend the scope of fire legislation to many existing buildings not designated under the Fire Precautions Act, in most cases it is to small workplaces. Emphasis is placed on self-compliance and responsibility is left to the employer, with the enforcing authority adopting a policing rather than a certifying role. The key regulation (Regulation 4) requires that a fire risk assessment is made of the workplace and that an emergency plan is prepared. The other regulations relate to the outcome of the assessment.

Hospitals and residential care premises were not designated under the Fire Precautions Act and so are covered by these regulations. Two draft guides were produced in the early 1980s outlining the standards which would be expected in these building if designation had occurred. The first of these covered hospitals, but was replaced in 1994 by Health Technical Memoranda 85 and 86 published by NHS Estates, with slightly modified Scottish versions (see section 8.5.4). The Home Office will accept that compliance with these documents is deemed to satisfy the Fire Precautions (Places of Work) Regulations. The second draft guide covers residential care premises: *Draft Guide to Fire Precautions in Existing Residential Care Premises* (Home Office, 1983). It outlines the standards which would have been required if these premises had been designated under the Fire Precautions Act. In Scotland this has been replaced with HTM 84 and in England and Wales a virtually identical document has been issued by the Institute of Building Control. However the Home Office has not yet withdrawn their guide, although it has no statutory force and is badly out of date.

The Fire Precautions (Places of Work) Regulations come into force in December 1997 and the enforcing authority for these regulations is again the fire authority. The regulations apply to England and Wales and to Scotland. Unfortunately, the regulations have been questioned by the European Commission on the grounds that they do not fully satisfy the requirements of the directives and at present (1998) the Home Office is considering how the regulations can be amended. The main criticism is over the exemptions given to premises already certified under the Fire Precautions Act, as the intention of the directives was to ensure continuous assessment and review, while obtaining a Fire Certificate means that the level is accepted as satisfactory. It is not clear how this dilemma is to be resolved.

8.2.5 Health and Safety at Work, etc., Act

The Health and Safety at Work, etc., Act of 1974 placed responsibility on employers, employees and others who may be affected by their actions for health, safety and welfare. Under this Act the Management of Health and Safety at Work Regulations came into force in January 1993. These regulations introduced into British law certain European directives on health and safety. The Act and its associated regulations are enforced by the Health and Safety Executive, and its Inspectorate.

Section 2 of the Health and Safety at Work, etc., Act places a duty of care on the employer towards the employees in respect of safety. The employer must provide information, training and supervision and also maintain the place of work as safe and provide means of access and egress that are safe. The employer is required to have a published safety policy and consult with safety representatives from the employees. Section 7 of the Act places a duty of care on employees for their own safety.

In the Regulations there is now a specific requirement to undertake risk assessments (Regulation 3) which is couched in very similar terms to the requirements of the Fire Precautions (Places of Work) Regulations. There is also a requirement (Regulation 7) to establish procedures for serious and imminent danger and danger areas, with fire being explicitly mentioned as one such danger. While it is unlikely that the Inspectorate would wish to become involved in specifying general building fire safety, as opposed to the safety of particular work processes, it is a potentially confused and contentious area. Designers will need to ensure that all statutory bodies make clear the reasons why they wish to become involved in a project and clarify the precise regulations under which they are working.

Under the Fire Certificates (Special Premises) Regulations 1976 (also made under the basic Act) certificates are required for such 'special premises' and these have very similar provisions and requirements to those issued under the Fire Precautions Act. There are also special requirements relating to the storage of hazardous substances (highly flammable liquids for example) and certain fire safety risks in manufacturing processes.

8.2.6 Licensing Acts

Under the Licensing Act of 1964, fire authorities can object to the issue of a licence to sell or supply intoxicating liquor or to refuse to register a club which supplies alcoholic drink if the means of escape are not satisfactory or there is undue risk of fire in the premises.

Under the Gaming Act of 1968, which covers casinos and bingo halls, fire authorities can object to the grant or renewal of a licence if the premises are unsuitable by reason of their layout, character, location, or if reasonable facilities to inspect the buildings have been refused.

In the case of both alcohol and gaming licences, application must be to the licensing court (licensing justices), who must consult the fire authority. The fire authority makes observations and the court makes a ruling which may include requirements to alter the premises. An applicant may appeal to the court within 14 days.

Under the Cinematograph Act of 1985 and the Theatre Act of 1968, local authorities are responsible for ensuring adequate fire safety in places of public entertainment before granting licences. The regulations cover means of escape, lighting and general fire precautions.

Although there is not prescriptive legislation as to the details of what can be required under the various Licensing Acts, a guide was published in 1990 which sets out suggested standards: *Guide to Fire Precautions in Existing Places of Entertainment and Like Premises* (Home Office/Scottish Home and Health Department, HMSO, 1990). This recommends standards for means of escape and other fire precautions in a wide range of premises used for public entertainment and recreation. It was drafted by a working party set up following the Stardust Disco fire in Dublin and is aimed mainly at Building Control Officers and Fire Prevention Officers. It aims to set acceptable standards and encourage consistency in enforcement.

8.2.7 Other legislation

Fire safety provisions are attached to a plethora of Acts often related to specific building types or specific activities, including:

*   animal boarding establishments
*   buildings for the disabled
*   caravan sites
*   children's homes
*   community care homes
*   educational buildings
*   explosives factories and workplaces
*   firework factories
*   mines and quarries
*   nurseries
*   petroleum installations
*   pipe lines

One particular area of concern is with 'Houses in Multiple Occupation' and here two documents may be of interest to the designer: Houses in Multiple Occupation, Guidance to Local Housing Authorities on Standards of Fitness under section 352 of the Housing Acts 1985, Department of the Environment Circular 12/92, and *Guide to Means of Escape and Related Fire Safety Measures in Existing Houses in Multiple Occupation in Scotland* (Scottish Office, 1988). Both these documents are currently under intensive review and there is also the possibility that the government will introduce the compulsory licensing of Houses in Multiple Occupation.

As well as the national legislation which has been mentioned, there is also a vast body of local legislation concerned with fire safety. Many counties have powers invested in them in respect of fire safety for different building types. Some local acts were introduced at the beginning of the century and many were transferred from the old authorities to their successors when local government re-organization occurred in 1974 and again in the 1990s. The government has committed itself to a process of repealing these Acts and incorporating their provisions within any new proposals for a fire safety bill. It is obviously nonsensical to apply different fire safety standards in different local government areas.

8.3 Northern Ireland legislation

Northern Ireland has its own legislative system both for new and existing buildings and therefore has to be considered completely separately.

New buildings are subject to the Building Regulations, which were revised and re-issued at the end of 1994. Sadly the opportunity to adopt the Approved Documents already in use with the English and Welsh Building Regulations was not taken. The fire section of the Northern Ireland Regulations is Part E, and this has eleven regulations. Regulations E2, E4, E6, E8 and E10 are functional standards which are identical with those in Part B in England and Wales. Regulations E3, E5, E7, E9 and E11 specify the deemed to satisfy documents relevant to each of the functional requirements. The first regulation (E1) is simply explanatory. The main document which is deemed to satisfy is the Technical Note published to accompany the regulations. This is very similar to Approved Document B to the English and Welsh Building Regulations and approximately 80% of it is identical. The variations which have been introduced confuse the issue and do not add either to the comprehensibility of the document or to its technical excellence.

Existing buildings are covered by the Northern Ireland equivalent of the Fire Precautions Act, the Fire Services (Northern Ireland) Orders of 1984 and 1993. These set out virtually identical conditions and standards to the Fire Precautions Act, but the list of designated premises is different. So far designations under this order are:

1. Leisure premises (1985)
2. Hotels and boarding-houses (1985)
3. Factory, offices and shop premises (1986)
4. Betting, gaming and amusement premises (1987).

Three guides were published in Northern Ireland to accompany the initial Order (Hotels and Boarding Houses, Factories, and Offices and Shops). However it is now accepted that compliance with the documents produced by the Home office will be deemed to satisfy the requirements of the Orders.

There are also regulations (1986) which outline the fire precautions to be taken in smaller factory, office and shop premises which do not require a fire certificate under the designation order.

Health and Safety issues are addressed through a series of Health and Safety Orders which mirror the provisions of the Health and Safety at Work, etc., Act and its associated regulations. The standards and requirements are intended to be identical with those in Great Britain. There is a separate Health and Safety Agency, with its own Inspectorate which includes specialist fire surveyors.

8.4 British Standards and International Standards

The British Standards Institution
389 Chiswick High Road, London W4 4AJ
0171 996 9000

British Standards are published by the British Standards Institution (BSI) and they are prefixed with the letters BS. International standards are published by the International Organization for Standardization and are prefixed by the letters ISO.

In addition to the standards listed below the BSI has published a *Draft for Development: Fire safety engineering in buildings,* 1997. This does not have the same status as a finished code but is extremely significant. It brings together a wealth of information on fire safety engineering and suggests a framework within which the fire safety engineering of a building can be designed.

The most significant standards for architects and designers in the field of fire safety are as follows:

BS 476: Fire Tests on Building Materials and Structures

Part 3: 1975 – External fire exposure roof test
This test assesses the ability of a roof structure to resist penetration by fire when the outer surface is exposed to radiation and flame, and the likely extent of surface ignition during penetration. Roof structures are classified according to the actual times recorded, for example, a 'P60' designation means that the sample of roof resisted penetration for at least 60 min. Roof classifications are prefixed by a designation showing whether it was tested as a sloping (S) or flat (F) construction.

Part 4: 1970 – Non-combustibility test for materials
This test assesses whether materials are non-combustible by seeing if samples will give off heat or flame when heated.

Part 5: 1979 – Method of test for ignitability
This test assesses whether a material will ignite when subjected to a flame for 10 s.

Part 6: 1989 – Method of test for fire propagation for products
This test assesses the contribution of combustible materials to fire growth when they are subjected to flame and radiant heat.

Part 7: 1987 – Method of classification of the surface spread of flame of products
This test assesses the spread of flame across flat materials (normally wall or ceiling linings) when they are subjected to flame and radiant heat. Materials are classified as follows:

Class	Max. flame spread at	
	1.5 min	10 min
1	165 mm	165 mm
2	215 mm	455 mm
3	265 mm	710 mm
4	900 mm	900 mm

Part 10: 1983 – Guide to the principles and application of fire testing
A basic guide.

Part 11: 1982 – Method for assessing the heat emission from building products
This test is a development of the basic test for non-combustibility in Part 4; and
it quantifies the level of heat given off by a material when it is heated.

Part 12: 1991 – Method of test for ignitability of products by direct flame
impingement
This test uses a choice of seven flaming ignition sources for a variety of flame
application times.

Part 13: 1987 – Method of measuring the ignitability of products subjected to
thermal irradiance
This test is a development of the basic test for ignitability in Part 5 and measures
the ease with which materials ignite when subjected to thermal radiation in the
presence of a pilot ignition source. It is identical to ISO 5657: 1986 – Fire tests.
Reaction to fire. Ignitability of building products

Part 15: 1993 – Method of measuring the rate of heat release of products

Part 20: 1987 – Method for determination of the fire resistance of elements of
construction (general principles)

Part 21: 1987 – Method for determination of the fire resistance of loadbearing
elements of construction

Part 22: 1987 – Method for determination of the fire resistance of non-loadbearing
elements of construction

Part 23: 1987 – Method for determination of the contribution of components to
the fire resistance of a structure
These tests assess the fire resistance of different elements of construction and
they replace Part 8 (1972 – Test methods and criteria for the fire resistance of
elements of building construction). The length of time for which the building
elements can satisfy the following criteria, under test conditions, is recorded:

– loadbearing capacity (supporting the test load without excessive deflection)
– integrity (resisting collapse, the formation of holes and the development of flaming on the unexposed face)
– insulation (resisting an excessive rise in temperature on the unexposed face).

Test results are normally expressed for each of the criteria in minutes. Columns and beams only have to satisfy the loadbearing criteria, glazed elements normally only have to satisfy the integrity criteria, while floors and walls have to satisfy all three criteria.

Part 24: 1987 – Method for determination of the fire resistance of ventilation ducts
This test assesses the ability of ductwork to resist the spread of fire from one compartment to another without the assistance of dampers. Results are given in minutes for each of the criteria of stability, integrity and insulation. It is identical to ISO 6944: 1985 – Fire resistance tests. Ventilation ducts

Section 31.1: 1983 – Methods for measuring smoke penetration through doorsets and shutter assemblies. Method of measurement under ambient conditions
Based on Part 1 (1981), the ambient temperature test, ISO 5925: Fire tests. Evaluation of performance of smoke control door assemblies. This test assesses the likely amount of smoke penetration through shut doorsets and shutter assemblies. Results are given in air leakage in cubic metres per hour.

Part 32: 1989 – Guide to full-scale fire tests within buildings

Part 33: 1993 – Full-scale room test for surface products

The British Standards Institution have also published the following documents on fire testing:
PD 6496: 1981 – A comparison between the technical requirements of BS 476 Part 8 (1972) and other relevant international standards and documents on fire-resistance tests.
PD 6520: 1988 – Guide to fire test methods for building materials and elements of construction

Other ISO standards relevant to testing which the architect might encounter include:
ISO 834: 1975 – Fire resistance tests. Elements of building construction
ISO 1182: 1983 – Fire tests. Building materials. Non-combustibility tests
ISO 1716: 1973 – Building materials. Determination of calorific potential
ISO 3008: 1976 – Fire resistance tests. Door and shutter assemblies
ISO 3009: 1976 – Fire resistance test. Glazed elements
ISO 3261: 1975 – Fire tests. Vocabulary

BS 750: 1984 – Specification for Underground Fire Hydrants and Surface Box Frames and Covers

BS 1635: 1990 – Graphic Symbols and Abbreviations for Fire Protection Drawings

BS 3169: 1986 – Specification for First-Aid Reel Hoses for Fire-Fighting Purposes

BS 4422: Glossary of Terms Associated with Fire
Part 1: 1987 – General terms and the phenomenon of fire
Part 2: 1990 – Building materials and structures
Part 3: 1990 – Fire detection and alarm
Part 5: 1989 – Smoke control
Part 6: 1988 – Evacuation and means of escape
Part 7: 1988 – Explosion detection and suppression means
This BS is gradually being revised so as to conform to ISO 8421: Fire protection. Vocabulary.

BS 4790: 1987 – Method of Determination of the Effects of a Small Ignition Source on Textile Floor Coverings (hot metal nut method)

BS 5041: Fire Hydrant Systems Equipment
Part 1: 1987 – Specification for landing valves for wet risers
Part 2: 1987 – Specification for landing valves for dry risers
Part 3: 1975 – Specification for inlet breechings for dry riser inlets
Part 4: 1975 – Specification for boxes for landing valves for dry risers
Part 5: 1974 – Specification for boxes for foam inlets and dry riser inlets

BS 5266: Emergency Lighting
Part 1: 1988 – Code of practice for the emergency lighting of premises other than cinemas and certain other premises used for entertainment

BS 5268: Code of Practice for the Structural Use of Timber
Section 4.1: 1978 – Method of calculating fire resistance of timber members
Section 4.2: 1989 – Recommendations for calculating fire resistance of timber stud walls and joisted floor constructions

BS 5306: Fire Extinguishing Installations and Equipment on Premises
Part 0: 1986 – Guide for the selection of installed systems and other fire equipment
Part 1: 1976 – Hydrant systems, hose reels and foam inlets
Part 2: 1990 – Sprinkler systems

Part 3: 1985 – Code of practice for selection, installation and maintenance of portable fire extinguishers
Part 4: 1986 – Specification for carbon dioxide systems
Section 5.1: 1992 – Halon 1301 total flooding systems
Section 5.2: 1984 – Halon 1211 total flooding systems
Section 6.1: 1988 – Specification for low expansion foam systems
Section 6.2: 1989 – Specification for medium and high expansion foam systems
Part 7: 1988 – Specification for powder systems

BS 5378: Safety Signs and Colours

Part 1: 1980 – Specification for colour and design
Part 2: 1980 – Specification for calorimetric and photometric properties of materials
Part 3: 1982 – Additional specifications

BS 5395: Stairs, Lobbies and Walkways

Part 1: 1977 – Code of practice for the design of straight stairs
Part 2: 1984 – Code of practice for design of helical and spiral stairs
Part 3: 1985 – Code of practice for the design of industrial type stairs, permanent ladders and walkways

BS 5445: Components of Automatic Fire Detection Systems

Part 1: 1977 – Introduction
Part 5: 1977 – Heat sensitive detectors – point detectors containing a static element
Part 7: 1984 – Specification for point type smoke detectors
Part 8: 1984 – Specification for high temperature heat detectors
Part 9: 1984 – Methods of test of sensitivity to fire

BS 5446: Specification for Components of Automatic Fire Alarm Systems for Residential Premises

Part 1: 1990 – Point-type smoke detectors

BS 5499: Fire Safety Signs, Notices and Graphic Symbols

Part 1: 1990 – Specification for fire safety signs
Part 2: 1986 – Specification for self-luminous fire safety signs
Part 3: 1990 – Specification for internally illuminated fire safety signs
(Note also ISO 6309: 1987 – Fire Protection. Safety signs)

BS 5588: Code of Practice for Fire Precautions in the Design of Buildings

Part 0: 1996 – Guide to fire safety codes of practice for particular premises/ applications
Part 1: 1990 – Code of practice for residential buildings
Part 4: 1998 – Code of practice for smoke control in protected routes using pressurization

Part 5: 1991 – Code of practice for fire fighting stairways and lifts
Part 6: 1991 – Code of practice for places of assembly
Part 7: 1997 – Code of practice for atrium buildings
Part 8: 1988 – Code of practice for means of escape for disabled people
Part 9: 1989 – Code of practice for ventilation and air conditioning ductwork
Part 10: 1991 – Code of practice for shopping malls
Part 11: 1997 – Code of practice for shops, offices, industrial, storage and other similar buildings

The British Standards Institution have also published the following documents to accompany BS 5588:

PD 6512 – Use of Elements of Structural Fire Protection with Particular Reference to the Recommendations given in BS 5588: Fire Precautions in the Design and Construction of Buildings
Part 1: 1985 – Guide to fire doors
Part 3: 1987 – Guide to the fire performance of glass

Note
BS 9999: Fire Precautions in Buildings
This set of documents is under development and is likely to replace all of BS 5588 plus many associated fire safety standards. It is intended to produce an integrated series of design guides covering all aspects of fire safety and is applicable to both new and existing buildings. It is hoped that these will eventually be used by the government departments as the technical guidance to support the Building Regulations and Technical Standards and to be cited by the Home Office in relation to a revised Fire Precautions Act. This is an ambitious project which will not be completed for a number of years, but which will greatly simplify and harmonize guidance and standards throughout the United Kingdom.

BS 5720: 1979 – Code of Practice for Mechanical Ventilation and Air Conditioning in Buildings

BS 5725: Emergency Exit Devices
Part 1: 1981 – Specification for panic bolts and panic latches mechanically operated by a horizontal pushbar.

BS 5839: Fire Detection and Alarm Systems in Buildings
Part 1: 1988 – Code of practice for system design, installation and servicing
Part 2: 1983 – Specification for manual call points
Part 3: 1988 – Specification for automatic release mechanisms for certain fire protection equipment
Part 4: 1988 – Specification for control and indicating equipment
Part 5: 1988 – Specification for optical beam smoke detectors

Part 6: 1995 – Code of practice for the design and installation of fire detection and alarm systems in dwellings
Part 8: 1998 – Code of practice for the design, installation and servicing of alarm systems

BS 5852: 1990 – Fire Tests for Furniture
Replaces BS 5852 Part 1 (1979) and BS 5852 Part 2 (1982) but these will still be referred to as they are cited in legislation.

BS 5950: Structural Use of Steelwork in Building
Part 8: 1990 – Code of practice for the fire protection of steelwork

BS 6266: 1992 – Code of Practice for Fire Protection for Electronic Data Processing Installations

BS 6336: 1982 – Guide to Development and Presentation of Fire Tests and their use in Hazard Assessment

BS 6387: 1994 – Specification for Performance Requirements of Cables Required to Maintain Circuit Integrity under Fire Conditions

BS 6535: Fire Extinguishing Media
Part 1: 1990 – Specification for carbon dioxide

BS 6575: 1985 – Specification for Fire Blankets

BS 6651: 1992 – Code of Practice for Protection of Structures against Lightning

BS 7176: 1995 – Specification for Resistance to Ignition of Upholstered Furniture

BS 7177: 1996 – Specification for Resistance to Ignition of Mattresses, Divans and Bed Bases

BS 7346: Components for Smoke and Heat Control Systems
Part 1: 1990 – Specification for natural smoke and heat exhaust ventilators
Part 2: 1990 – Specification for powered smoke and heat ventilators
Part 3: 1990 – Specification for smoke curtains

BS 8110: Structural Use of Concrete
Part 1: 1997 – Code of practice for design and construction
Part 2: 1985 – Code of practice for special circumstances
Part 3: 1985 – Design charts to accompany Part 1

BS 8202: Coatings for Fire Protection of Building Elements
Part 1: 1995 – Code of practice for the selection and installation of sprayed mineral coatings
Part 2: 1992 – Code of practice for the use of intumescent coating systems

BS 8214: 1990 – Code of Practice for Fire Door Assemblies with Non-Metallic Leaves
Recommendations for the specification, design, construction, installation and maintenance of fire door assemblies.

8.5 Guidance

The amount of guidance available to the design team is deceptively extensive. There are a large number of government organizations, trade associations and private companies offering information, guidance and advice on fire safety; however, this information is of variable quality and extremely patchy in its coverage. Some areas are very well covered (e.g. auto-suppression systems), while in other fields (e.g. smoke control) there is only a scattering of experts and information usable by the design team. One of the problems facing the architect is establishing what information is reliable and trustworthy; and where there is a need to take further more specialist advice. This section cannot hope to cover all the guidance available, but it does seek to consider the main sources of information and to identify the most recent and useful publications from these different sources. Where the title of documents is not completely self-explanatory an additional note has been added to guide designers as to the relevance and application of the materials.

The guidance has been listed under the organization producing it as this is the most likely way that designers will be able to trace what they need. It would have been possible to organize materials by reference to the fire safety tactic to which they relate, but this would have resulted in considerable duplication. The useful life of a code or guide is only about ten years, and therefore there is a continually shifting body of information. However as specific documents go out of date they are likely to be replaced by the organization concerned, another reason for structuring this section by organization. Addresses are also included wherever possible to enable practices to obtain their own copies of key documents.

8.5.1 Fire Research Station

Fire Research Station
Building Research Establishment
Bucknalls Lane, Garston, Watford, Hertfordshire WD2 7JR
01923–664000

The Fire Research Station (FRS) is part of the Building Research Establishment and has been involved in fire safety research for over 40 years. There is a staff of some 150 who are involved in both fire research and the fire testing of buildings and products. There are facilities for full-scale fire tests, including an old airship hanger at Cardington which is capable of taking reconstructions of complete buildings.

The FRS is of value to architects both through its library and consultancy services, and because of its numerous publications. Many of these publications are research reports of limited value to architects in practice, but certain of their publications are useful reference documents and the designer may well find them being referred to by the statutory authorities when assessing designs. Some are full-scale books or reports, and others shorter works (digests) or single sheet information papers. The following list is far from comprehensive, but does identify the ones liable to be of some use in the design process.

Books, Reports and Digests

1 *Psychological Aspects of Informative Warning Systems*, D Canter *et al.*, BR127, 1988.
   Some general guidance useful for architects involved in the specification of communications systems.
2 *Guidelines for the Construction of Fire-resisting Structural Elements*, WA Morris *et al.*, BR128, 1988.
   Contains a set of tables of notional periods of fire resistance for structural elements based on current test data and information; it also includes a valuable summary of general fire safety information on the principal materials and elements of construction and the tables cover:
• Masonry construction (solid, hollow and cavity walls)
• Timber framed internal walls (loadbearing and non-loadbearing)
• Timber or steel framed external walls (loadbearing and non-loadbearing)
• Concrete columns, beams, and floors (both plain soffit and ribbed open soffit)
• Encased steel columns and beams
• Timber floors
3 *Fire Performance of External Thermal Insulation for Walls of Multi-storey Buildings*, BFW Rogowski *et al.*, BR135, 1988.
   Description of experimental tests, leading to fundamental design recommendations on external thermal insulation. Tables of design recommendations for both sheeted and non-sheeted systems.
4 *Increasing the Fire Resistance of Existing Timber Floors*, BRE Digest 208 (new edition), 1988.
   Explains how periods of fire resistance of up to 1 hour may be obtained by upgrading existing timber floors. It covers the addition of protection to the underside of the ceiling, over the floorboarding and between the joists.
5 *Design Principles for Smoke Ventilation in Enclosed Shopping Centres*, HP Morgan and JP Gardner, BR186, 1990.

The most useful document on smoke ventilation available for designers. It is an update on the 1979 FRS publication on the same subject and is the result of work carried out by FRS and Colt International Limited. It is an invaluable explanation of basic smoke flow calculations and their application to large volume spaces. Nice, easily understood drawings and not too many unnecessary calculations.

6 *Experimental Programme to Investigate Informative Fire Warning Characteristics for Motivating Fast Evacuation*, L Bellamy and T Geyer, BR 172, 1990.
Research report on innovative forms of alarm systems, some limited design use to architects involved in major developments.

7 *External Fire Spread: building separation and boundary distances*, BR 187, 1991.
Describes the different methods for calculating adequate space separation between buildings. Prepared to support Approved Document B.

8 *Fire Modelling*, BRE Digest 367, 1991.
A very short paper outlining a proposed fire model. Mixture of generalized statements and very precise calculations (of little value to designers).

9 *Sprinkler Operation and the Effect of Venting: studies using a zone model*, P Hinkley, BR213, 1992.
Describes a mathematical model developed by FRS and Colt International Ltd to assess the interaction between sprinklers and smoke venting. Intended for the fire engineer rather than the architect.

10 *Aspects of Fire Precautions in Buildings*, REH Read and WA Morris (3rd Edition), 1993.
An extremely useful guide to certain aspects of passive fire precautions, particularly means of escape and structural fire protection. Unfortunately it does not provide a comprehensive guide for designers; it has good sections on standard fire tests and also extensive coverage of the history of fire safety legislation.

11 *Design Approaches for Smoke Control in Atrium Buildings*, GO Hansell and HP Morgan, BR258, 1994.
Excellent introduction to the principles for the design of smoke control systems.

12 *Escape of Disabled People from Fire: a measurement of and classification of capability for assessing escape risk*, TJ Shields *et al.*, BR301, 1996.
Provides some useful guidance on the problems of designing for and managing the escape of disabled people from buildings.

Selected Information Papers

Selection of Sprinklers for High Rack Storage in Warehouses, IP 5/88.
Thermal Bowing in Fire and how it affects Building Design, IP 21/88.
The Development of a Fire Risk Assessment Model, IP 8/92.
False Alarms from Automatic Fire Detection Systems, IP 13/92.

8.5.2 Loss Prevention Council

Loss Prevention Council
Melrose Avenue, Borehamwood, Hertfordshire, WD6 2BJ
0181 207 2345

The Loss Prevention Council (LPC) is involved in all aspects of loss prevention and control, including fire safety. It is funded by the Association of British Insurers and Lloyd's and has a number of component parts. The first is the LPC Laboratories, which has comprehensive fire testing facilities and is also involved in establishing new standards. The second part is the Loss Prevention Certification Board, which operates a number of certification schemes for products. The third component of the LPC is possibly the most important for architects, the Fire Protection Association (FPA). The FPA provides information and advice on all aspects of fire safety, including fire safety design. It produces a number of valuable publications and an authoritative journal reviewing recent fires.

The final component part of the LPC is the National Approval Council for Security Systems, which is based separately at Maidenhead and is more concerned with security than with fire.

The first general publication is a List of Approved Products and Services, which is produced each year and provided free on request both as a book and as a CD-ROM. Other publications are as follows:

LPC Rules (formerly published by the Fire Offices' Committee) and Codes of Practice give fundamental information related to loss prevention measures:

1. LPC Rules for Automatic Fire Detection and Alarm Installations for the Protection of Property, 1991
   Design rules for fire alarm installations to protect property.
2. LPC Rules for Automatic Sprinkler Installations, 1994
   This is the only document to include both the British Standards (BS 5306 Part 2) and the insurers' additional requirements. Technical Bulletins to accompany these LPC Rules include:
   – Sprinkler Systems for Dwelling Houses, Flats and Transportable Homes
   – Sprinkler Protection of Intensive Hanging Garment Stores
   – Automatic Sprinkler Pump Testing and Commissioning
   – Sprinkler Head Design Characteristics
   – Supplementary Requirements for Sprinkler Installations Operating in the Dry Mode
   – Automatic Sprinkler Protection to High Rise and Multiple Storey Buildings
3. LPC Design Guide for Fire Protection of Buildings, 1997
   This replaces their Code of Practice for the Construction of Buildings and

contains valuable information for architects and designers which will help them comply with likely insurance requirements. It is a loose-leaf document accumulating work currently available.
4. Fire Prevention on Construction Sites, 4th Edition, 1997
   A short guide which should be required reading for all contractors.

LPC Recommendations (formerly published by the Fire Offices' Committee), gives practices which insurers expect to be followed during the building process and use of the building. These include:

1. Fire Protection of Atrium Buildings, 1990.
2. Loss Prevention in Electronic Data Processing and Similar Installations, Part 1 Fire Protection, 1990.
3. Fire Protection of Laboratories, 1991.

LPC Standards, Quality Schedules and Related Documents give details of particular standards for different components and systems. These are mainly concerned with communication systems, extinguishing systems, fire doors, glazing and fire protection to steelwork.

The LPC Library of Fire Safety brings together advisory information on all aspects of fire, its prevention and control. The compendium consists of six handbooks. These are revised and added on a regular basis.

1. *Fire Protection Yearbook*, 1998/9
2. *Fire and Hazardous Substances*, 1994
3. *Guide to Building Fire Protection*, 1997
4. *Guide to Fire Safety Signs*, 1997
5. *Workplace Fire Risk Management: A Guide for Employers*, 1997
6. *Fire Safety Management and Training*, 1998.

Obviously Volume 3 is of most interest to the design team, however any large architectural practice should have the complete set within its library.

The LPC has also published two useful documents on fire safety in historic buildings:

1. Heritage Under Fire – a guide to the protection of historic buildings, Second Edition, 1995 (with the UK Working Party on Fire Safety in Historic Buildings).
2. The Fire Protection of Old Buildings and Historic Town Centres, 1992.

8.5.3 London District Surveyors Association

Until the introduction of the 1985 Building Regulations, Inner London had its own set of building bye-laws administered by District Surveyors. The London District Surveyors Association remains in existence and has produced three significant documents. The London Fire and Civil Defence Authority played a prominent role in the production of these three guides and they are therefore used as the basis for the appraisal of applications for approval as part of the formal consultation procedures. The three guides are:

1. Fire Safety Guide No 1 – Fire Safety in Section 20 Buildings, 1990.
   This is based on a previous Greater London Council Document, which has now been amended and rewritten to harmonize with the 1985 Building Regulations.
2. Fire Safety Guide No 2 – Fire Safety in Atrium Buildings, 1989.
3. Fire Safety Guide No 3 – Phased Evacuation from Office Buildings, 1990.

8.5.4 NHS Estates, Department of Health

NHS Estates, an executive agency of the Department of Health, have produced a whole series of guidance documents (known as FIRECODE) on the fire safety of healthcare premises. Compliance with FIRECODE is required by the Secretary of State for Health and certain of the documents are also recognized by the Department of the Environment as Approved Documents, or by the Home Office as prescribing the standards of fire safety required in the workplace. The documents are for use in England and Wales and in Northern Ireland (except where stated) and are available from HMSO. In Scotland a slightly modified set of documents was produced in 1998 under the title 'NHS in Scotland Firecode' and these are only available on a CD-ROM produced by the NHS in Scotland Environmental Forum.

Policies and Principles
This contains basic policy, principles and key management guidance. It is a short document intended for health service managers. Separate England and Wales, and Northern Ireland versions were both published in 1994. No Scottish equivalent.

Fire Precautions in New Hospitals, Health Technical Memorandum (HTM) 81, 1996
This sets out the fire safety requirements for all new hospitals. Scottish equivalent on CD-ROM.

Fire Risk Assessment in Nucleus Hospitals, 1997
This sets out the fire safety assessment methods for all hospitals built under the

NUCLEUS system used during the 1980s and early 1990s. Nucleus was never used in Scotland.

Alarm and Detection Systems, HTM 82, 1996
This provides specifications for alarm and detection systems for both new hospitals and for the upgrading of existing hospitals. Scottish equivalent on CD-ROM.

Fire Safety in Healthcare Premises: General Fire Precautions, HTM 83, 1994
This provides guidance for fire safety officers within health care buildings and so is of less relevance to designers. Scottish equivalent on CD-ROM.

Fire Safety in Residential Care Premises, HTM 84, 1995
This provides guidance on the standards for both new and existing residential care premises. It only applies in Northern Ireland. Scottish equivalent on CD-ROM. A version for England and Wales was published by the Institute of Building Control in 1997, but this is only advisory and has no statutory force or government endorsement.

Fire Precautions in Existing Hospitals, HTM 85, 1994
This lays down the standard of fire safety required in existing hospitals. Scottish equivalent on CD-ROM.

Fire Risk Assessment in Hospitals, HTM 86, 1994
This outlines a method of assessing the fire safety in existing premises against the standards laid down in HTM 85. It is one of the very few systematic methods available to assess fire safety on a true fire engineering basis. Scottish equivalent on CD-ROM.

Textiles and Furniture, HTM 87, 1993
This provides guidance on the specification of furniture, fabrics and fittings for healthcare premises. This only applies in England and Wales. Scottish equivalent on CD-ROM.

Guide to Fire Precautions in NHS Housing in the Community for Mentally Handicapped and Mentally Ill People, HTM 88, 1986
This provides guidance on fire precautions for new purpose built buildings for the care of mentally handicapped and mentally ill patients within the community. This only applies in England and Wales.

Escape Bed Lifts, FPN 3, 1987
Scottish equivalent on CD-ROM.

Hospital Main Kitchens, FPN 4, 1994

Scottish equivalent on CD-ROM.

Commercial Enterprises on Hospital Premises, FPN 5, 1992
Scottish equivalent on CD-ROM.

Arson Prevention and Control in Healthcare Premises, FPN 6, 1994
Scottish equivalent on CD-ROM.

Fire Precautions in Patient Hotels, FPN 7, 1995
Scottish equivalent on CD-ROM.

NHS Healthcare Fire Statistics 1994/5, FPN 9, 1996

Laboratories on Hospital Premises, FPN 10, 1996
Scottish equivalent on CD-ROM.

8.5.5 Department of Education

The Department of Education have published within their building bulletins series an extremely good guide to fire safety in educational buildings. This is:

Fire and the Design of Educational Buildings, Building Bulletin 7, HMSO, 1988
Building Bulletin 7 was first published in 1952, the current sixth edition has been updated to complement the Approved Documents issued with the 1985 English and Welsh Building Regulations. It has a legal status and in effect provides the code of practice for fire safety in educational buildings. It is primarily aimed at new design work, but may also be appropriate in the context of alterations to existing buildings. Fortunately it has an excellent "first principles" approach and is clearly illustrated and explained. It covers:

* Means of escape
* Precautions against fire
* Structural fire precautions
* Fire warning systems and fire fighting
* Fire safety management issues within the buildings.

8.5.6 Timber Research and Development Association

Timber Research and Development Association
Stocking Lane, Hughenden Valley, High Wycombe, Buckinghamshire
HP14 4ND
01494 563091

The Timber Research and Development Association (TRADA) is a research and testing organization for the timber industry. It produces a series of valuable publications for architects, the Wood Information Sheets (WIS); and provides test certificates for certain products.

1. *Flame Retardant Treatments for Timber*, WIS 2/3–3, 1988 under revision 1992
   Describes surface finishes and impregnation treatments for timber and wood based sheet materials; and relates these to the relevant British Standards.
2. *Low Flame Spread Wood-based Board Products*, WIS2/3–7,1987 under revision 1992
   Describes products which have treatments applied to the particles, veneers or fibres during or after manufacture so that the board has low surface spread of flame properties.
3. *Timber and Wood Based Sheet Materials in Fire*, WIS 4–11, revised 1991
   Describes the behaviour of timber in fire.
4. *Technology of Fire Resisting Doorsets*, WIS 1–13, 1989
   Describes the performance requirements for fire resisting doorsets, and gives guidance on door furniture and glazing.
5. *Fire Resisting Doorsets by Upgrading*, WIS 1–32, revised 1991
   Gives guidance on the assessment of existing doorsets for upgrading; and on the techniques of upgrading.
6. *Timber Building Elements of Proven Fire Resistance*, WIS 1–11
   A series of sheets giving details of timber constructions which have had their fire resistance validated by test.
   D5 to D12: doors, revised 1991
   M1 to M4: glazed screens, 1983 and 1987.

8.5.7 Steel Construction Institute

Steel Construction Institute
Silwood Park, Ascot, Berkshire, SL5 7QN
01344 623345

The Steel Construction Institute (SCI) is a research and development organization for the steel construction industry. It has produced a number of useful detailed documents on the fire protection of steelwork, as well as a number of Technical Reports giving test data and an excellent fire report.

1. *Fire and Steel Construction: The Behaviour of Steel Portal Frames in Boundary Conditions*, GM Newman, 2nd Edition, 1990
   Describes the behaviour of portal frames in fire, and gives guidance on the design of column bases to resist rafter collapse.
2. *Fire Resistant Design of Steel Structures – A Handbook to BS 5950 Part 8*, RM Lawson and GM Newman, 1990

A user friendly guide to the British Standard with examples of the evaluation of fire resistance.

3. *The Fire Resistance of Composite Floors with Steel Decking*, GM Newman, 2nd Edition, 1991
Outlines a fire engineering method of verifying fire resistance.

4. *Technical Report: Fire Resistance of Composite Beams*, GM Newman and RM Lawson, 1991

5. *Investigation of Broadgate Phase 8 Fire*, SCI, 1991

6. *Fire Protection for Structural Steel in Buildings*, 2nd Edition, 1992
Compilation of techniques for fire protection. Published by the Association for Specialist Fire Protection with the Steel Construction Institute and Fire Test Study Group.

7. *Technical Report: Enhancement of Fire Resistance of Beams by Beam to Column Connections*, RM Lawson, 2nd Edition, 1992

8. *Technical Report: The Fire Resistance of Web-filled Steel Columns*, GM Newman, 1992

9. *Technical Report: The Fire Resistance of Shelf Angle Floor Beams to BS 5950: Part 8*, GM Newman, 1993

10. *Building Design Using Cold Formed Steel Sections: Fire Protection*, RM Lawson, 1993

11. *Structural Fire Design to EC3 and EC4, and Comparison with BS 5950*, RM Lawson and GM Newman, 1996

12. *Structural Fire Design: Off-site applied thin film intumescent coatings*, Ed. E Yandizo *et al.*, 1996

8.5.8 Textbooks

There are surprisingly few textbooks on the fire safety of buildings and anything published more than ten years ago will already be out of date in terms of legislation, building types and materials. The nine listed below are those which may be of some value to the architect and the design team, although only certain parts of each are relevant.

1 *Buildings and Fire*, TJ Shields and GW Silcock, 1989, Longman Scientific and Technical
A very technical book most suitable for the specialist fire safety engineer and probably too detailed for the average architect.

2 *Underdown's Practical Fire Precautions*, 3rd Edition revised by R Hirst, 1989, Gower
A reference handbook aimed at fire engineers rather than designers. Particularly useful for the sections on the technology of active fire fighting systems.

3 *Fire Safety and the Law*, J Holyoak, A Everton, and D Allen, Paramount Publishing, 2nd Edition, 1990

As the subtitle describes it, this is 'a guide to the legal principles related to the hazard of fire'. This is a legal rather than a technical book, but the chapters on current legislation and professional liabilities would be of interest to designers.

4 *Fire Safety in Tall Buildings*, Council on Tall Buildings and Urban Habitat, McGraw-Hill, 1992

A collection of papers produced to deal with different problems related to this building type. Too technical, unless you have to solve the problems of such buildings.

5 *Croner's Guide to Fire Safety*, Colin Todd, Croner, 1992

A general non-specialist textbook which covers similar issues to those addressed in this book in approximately similar depth.

6 *Design Against Fire*, (eds) Paul Stollard and Lawrence Johnston, E & FN Spon, 1993.

This is based on the keynote papers in an innovative postgraduate fire engineering course held at Queen's University of Belfast. It presents the collected views of the leading experts in fire safety engineering including John Abrahams, Jack Anderson, Dougal Drysdale, Lawrence Johnston, Bill Malhotra, Howard Morgan, John Northey, Jonathan Sime and Paul Stollard.

7 *Design for Fire Safety*, Paul Stollard and Lawrence Johnston, Emap Construct, 1995

A simple guide for those considering fire safety engineering alternatives to Approved Document B, providing a first point of reference rather than an authoritative resource (for that see the next textbook).

8 *SFPE Handbook of Fire Protection Engineering*, SFPE and NFPA, 2nd Edition, 1995.

This expensive American publication from the Society of Fire Protection Engineers (SFPE) and the National Fire Protection Association (NFPA) includes over sixty specialist articles covering the whole range of fire safety engineering. It is a principal reference book for all fire safety engineers, though its size, complexity and cost mean that it is probably not a first choice text for architects.

9 *Fire Engineering*, CIBSE Guide E, 1997

This publication from the Chartered Institute of Building Services Engineers (CIBSE) provides the engineering relationships required for the fire safety design of buildings. Another useful reference intended for the fire safety engineer rather than the architect, but written in a very accessible style.

8.6 Consultancy and advisory services

The previous sections have all dealt with written guidance materials, but it may be necessary for the design team to seek direct help from a fire safety consultancy. In this case there are a number of options for the design team to consider. The FRS and LPC have already been mentioned as providing advisory services on a consultancy basis.

The Fire Research Station offers two levels of consultancy. At its simplest their library services will assist architects and designers to find the correct documents and guidance from among both their own and any other publications. They also offer a technical consultancy which will advise directly on the fire engineering of a particular design. This might range from the complete integrated smoke control system, to the likely performance of a single component or assembly. They have an expert staff and extremely good test facilities for experimental work.

The different constituent parts of the Loss Prevention Council offer different services. The FPA has an information service which provides free advice and guidance to member organizations; the LPC Technical Centre offers full facilities for the fire testing of building and assemblies. The Loss Prevention Certification Board will give information on products which have been certified under the schemes.

Most architects who seek fire safety advice will approach one of the independent fire safety consultancies now offering this service. As with all consultants, some are excellent and some are little more than charlatans. It is essential for the architect to decide what services they need and therefore the right consultancy to provide such services. It may be that the architects are only seeking someone to check their own designs for legislative compliance, or they may need to use chartered fire engineers who would be capable of the much more demanding task of preparing a full fire engineering solution to an unusual design problem. As has been outlined in Chapter 7 the principal organizations have merged over the last few years and this has enabled them to gain recognition from the Engineering Council. Therefore designers wishing to appoint a chartered fire engineer should approach the Engineering Council Division of the Institution of Fire Engineers (IFE). The IFE offices are at: 148 New Walk, Leicester, LE1 7QB (0116 255 3654).

Although trade associations are established to serve the interests of their members and therefore may not always offer completely independent advice, they can provide a useful source of information to designers. This is particularly true when it comes to the selection of materials or components to meet an already determined specification. Trade associations in the field of fire safety which architects might wish to consult are as follows:

Association for Specialist Fire Protection
235 Ash Road, Aldershot, Hampshire, GU12 4DD
01252 321322

British Fire Protection Systems Association
Neville House, 55 Eden Street, Kingston Upon Thames, Surrey KT1 1BW
0181 549 5855

European Federation of Fire Separating Element Producers
20 Park Street, Princes Risborough, Buckinghamshire HP27 9AH
01844 275500

Fire Extinguishing Trades Association
Neville House, 55 Eden Street, Kingston Upon Thames, Surrey, KT1 1BW
0181 549 8839

Glass and Glazing Federation
44/48 Borough High Street, London SE1
0171 403 7177

Intumescent Fire Seals Association
20 Park Street, Princes Risborough, Buckinghamshire HP27 9AH
01844 275500

Lastly it is important to include CERTIFIRE. This is the certification authority
for passive fire protection products and services.

CERTIFIRE
101 Marshgate Lane, London E15 2NQ
0181 555 3234.

8.7 Glossary of fire terms

acceptability the extent to which the fire safety meets that deemed by society to
be necessary. This may be expressed through legislation and influenced by the
frequency of disasters.

active containment measures to contain the spread of fire which require the
operation of some form of mechanical device (e.g. the operation of smoke
vents or the release of fire shutters).

addressable system a comprehension/analysis system which compares the
present situation with data stored in the system's memory, to derive more
than just fault and fire signals from the detectors; and to permit the precise
location of the fire to be established.

alternating sprinkler system a sprinkler system which can be changed from
'wet' operation in summer to 'dry' operation in winter.

appraisal/assessment an estimate of the fire risks and fire precautions within a
building.

audit an examination of the appraisal/assessment to check for accuracy in the
estimation of fire risks; and the appropriateness and implementation of the fire
precautions.

auto-suppression fire extinguishing systems which activate automatically on
detecting a fire.

bridgeheads intended safe bases for firefighters attempting to tackle a fire within the building, served by protected lifts and adjacent to rising mains.

burning brands flaming debris carried by convection currents from buildings on fire.

combustibility the ease with which a material will burn when subjected to heat from an already existing fire.

combustion exothermic reaction of a combustible substance with an oxidizer.

communication the fire safety tactic of ensuring that if ignition occurs the occupants are informed immediately and any active fire systems are triggered.

compartmentation the technique of dividing a building into a number of compartments or subcompartments.

compartments fire- and smoke-tight areas into which a building can be divided to contain fire growth and limit travel distance, offering at least 60 min resistance.

compartment floor a fire-resisting floor used to separate one fire compartment from another and having a minimum period of resistance of 60 min.

compartment wall a fire-resisting wall used to separate one fire compartment from another and having a minimum period of resistance of 60 min.

conduction the transfer of heat by direct physical contact between solids.

containment the fire safety tactic of ensuring that the fire is contained in the smallest possible area, limiting the amount of property likely to be damaged and the threat to life safety.

convection the transfer of heat by the movement of the medium in liquids and gases.

conventional system a comprehension/analysis system where it is only possible to establish fault or fire signals from the detectors.

diffusion flames flames from combustion, where the rate of burning is determined by the rate of mixing of flammable vapours from a solid or liquid fuel and oxygen.

dry rising main a rising main normally kept empty of water, but to which the fire service can supply water at ground level in the event of a fire.

dry sprinkler system a sprinkler system where the majority of pipework is air filled until it is triggered.

egress direct escape to the outside.

emergency lighting lighting provided by standby generators on the failure of the mains supply.

envelope protection the limitation of the threat posed by a fire to adjoining properties and to people outside the building; and the threat posed by a fire in an adjoining property.

equivalency the provision of the same level of fire safety by different combinations of fire safety measures.

escape the fire safety tactic of ensuring that the occupants of the building and the surrounding areas are able to move to places of safety before they are threatened by heat and smoke.

escape lighting lighting provided on the failure of the normal lighting circuits by self-contained fittings.

extinguishment the fire safety tactic of ensuring that the fire can be extinguished quickly and with minimum consequential damage to the building.

final exit the termination of an escape route from a building giving direct access to a place of safety outside the building.

fire

1 self-supporting combustion characterized by the emission of heat and effluent often accompanied by flame and/or glowing combustion.

2 combustion spreading uncontrolled in time or space.

fire door a door or shutter provided for the passage of persons, air or objects which, together with its frame and furniture as installed in a building, is intended when closed to resist the passage of fire and/or gaseous products of combustion and is capable of meeting specified performance criteria to those ends.

fire engineering design which considers the building as a complex system, and fire safety as one of the many interrelated subsystems which can be achieved through a variety of equivalent strategies.

fire growth curve the relationship between the time from ignition and the size of the fire.

fire hazard the potential for injury and/or damage from fire.

fire load the quantity of heat which could be released by the complete combustion of all the combustible materials in a volume, including the facings of all bonding surfaces.

fire point the minimum temperature at which material ignites and continues to burn for a specified time after a standardized small flame has been applied to its surface under specified conditions.

fire precautions measures which can be taken to reduce the likelihood of ignition occurring and/or mitigate the consequences should ignition occur, including the fire safety tactics of prevention, communication, escape, containment and extinguishment.

fire propagation the degree to which a material will contribute to the spread of a fire through heat release, when it is itself heated.

fire protection measures to limit the effects of fire, including the fire safety tactics of communication, escape, containment and extinguishment.

fire resistance the ability of an item to fulfil for a stated period of time the required stability and/or integrity and/or thermal insulation and/or other expected duty specified in a standard fire-resistance test.

fire risk the product of:

1 probability of occurrence of a fire to be expected in a given state, and

2 consequence or extent of damage to be expected on occurrence of a fire.

fire safety components the specific building elements, structures and procedures, which the architect can use tactically to achieve fire safety – e.g. they may include fire doors, sprinklers, escape stairs and fire drills.

fire safety objectives the objectives which the architect must satisfy in order to achieve a fire-safe building, normally life safety and property protection.

fire safety tactics the tactics which the architect can adopt in order to satisfy the fire safety objectives, normally prevention, communications, escape, containment and extinguishment.

fire stop a seal provided to close an imperfection of fit or design tolerance between elements or components, to restrict the passage of fire and smoke.

flame zone of combustion in the gaseous phase usually with the emission of light.

flammable limits the range of concentrations of flammable vapours to air within which a flame will be produced in the presence of an ignition source.

flashover the transition to a state of total surface involvement in a fire of combustible materials within an enclosure.

flash point the minimum temperature to which a material or product must be heated for the vapours emitted to ignite momentarily in the presence of a flame under specified test conditions.

fuel limitation the control of the amount of fuel within a building or room.

fuel load the amount of potential fuel within a building or room, including both the building's fabric and contents.

ignitability measure of the ease with which a material can be ignited, under specified conditions.

ignition initiation of combustion.

ignition prevention measures taken to reduce the probability of ignition.

ignition risk the probability of ignition.

ignition source a source of energy which initiates combustion.

insulation the resistance offered by a material to the transfer of heat.

integrity the ability of a separating element when exposed to fire on one side, to prevent the passage of flames and hot gases or the occurrence of flames on the unexposed side, for a stated period of time in a standard fire-resistance test.

intumescents materials which react to heat by expanding and forming an insulating layer.

ionization detector a smoke detector which can identify the reduction in electrical current across an air gap in the presence of a small radioactive source, when smoke particles are present.

laminated glass glass which incorporates layers of transparent and translucent intumescent material.

loadbearing capacity the dimensional stability of a material.

manual call point (or **break glass call point**) an alarm switch which can be activated by the occupants.

neutral plane the level within a building where the internal air pressure is equivalent to atmospheric pressure.

occupancy load factor a measure for calculating the likely number of people in particular building types for a given floor area.

optical detector a smoke detector which can identify the reduction in light received from a light source by a photo-electric cell when smoke particles are present.

passive containment measures to contain the spread of fire which are always present and do not require the operation of any form of mechanical device – e.g. the fire protection provided to the building structure or the fire-resisting walls provided to divide a building into different compartments.

phased evacuation the planned evacuation of a building in stages.

place of safety an area to which the occupants can move, where they are in no danger from fire.

pre-action sprinkler systems one which is 'dry', but where water is allowed in on a signal from a more responsive detector (usually smoke) in advance of the heads being triggered.

pre-mixed flames flames from combustion where the fuel is a gas and is already mixed with oxygen.

pressurization the technique whereby staircases or corridors are pressurized so that they can resist the inflow of smoke.

prevention the fire safety tactic of ensuring that fires do not start by controlling ignition and limiting fuel sources.

progressive horizontal evacuation evacuation of the occupants away from a fire into a fire-free compartment or subcompartment on the same level.

protected shaft a shaft which enables persons, air or objects to pass from one compartment to another; and which is closed with fire-resisting construction.

radiation the transfer of heat without an intervening medium between the source and the receivers.

re-cycling (or **re-setting**) **sprinkler system** a sprinkler system where the heads can be closed once the fire is extinguished so that water damage is minimized.

refuge a place of safety within a building.

re-setting (or **re-cycling**) **sprinkler system** a sprinkler system where the heads can be closed once the fire is extinguished so that water damage is minimized.

rising mains vertical pipes installed within tall buildings (usually over 18 m), which have a fire service connection or booster pump at the lower end and outlets at different levels within the building.

sacrificial timber the technique of deliberately oversizing timber elements to enhance their fire resistance.

smoke visible part of the fire effluent.

smoke curtains barriers which can restrict the movement of smoke.

smoke layering (or **smoke stratification**) the process whereby different layers or zones develop within smoke due to the buoyancy of the gases involved.

smoke load the amount of potential fuel within a building or room which will produce smoke.

smoke obscuration reduction in visibility due to smoke.

smoke reservoirs areas in the ceiling of a space in which smoke will collect.

smoke stratification (or **smoke layering**) the process whereby different layers or zones develop within smoke due to the buoyancy of the gases involved.

smoke venting the technique of allowing smoke to escape from a building, or of forcing it out by mechanical means.

soot fine particles, mainly carbon, produced and deposited during the incomplete combustion of organic materials.

spontaneous ignition temperature the minimum temperature at which ignition is obtained under specified test conditions without any source of pilot ignition.

sprinklers auto-suppression systems using water sprays to extinguish small fires and contain growing fires until the fire services arrive.

stage 1 escape escape from the room or area of origin.

stage 2 escape escape from the compartment or subcompartment of origin by the circulation route to a final exit, a protected stair, or an adjoining compartment offering refuge.

stage 3 escape escape from the floor of origin to ground level.

stage 4 escape final escape at ground level.

structural elements loadbearing elements of the building, in particular floors and their supporting structures – e.g. columns or loadbearing walls.

structural protection fire resistance provided to structural elements.

subcompartments fire- and smoke-tight areas into which a building can be divided to reduce travel distances, offering 30 min of resistance.

subcompartment wall a fire-resisting wall used to separate one sub-compartment from another and having a minimum period of resistance of 30 min.

surface spread of flame the propagation of flame away from the source of ignition across the surface of a liquid or solid.

toughened glass glass which has been toughened to achieve higher levels of stability and integrity.

trade-offs the technique of providing equivalent levels of fire safety through different fire safety measures.

traditional fire safety design design which considers the building as a series of components and attempts to achieve fire safety by ensuring that all such components meet a specified performance standard.

travel distance the distance to be travelled from any point in a building to a place of safety, often specified in terms of the different stages of escape.

triangle of fire a description of the three ingredients necessary for a fire (heat, fuel and oxygen).

unprotected areas unprotected areas in relation to a side or external wall of a building means:

1 a window, door or other opening,

2 any part of the external wall which has a fire resistance less than 60 min or 30 min if single storey (integrity and loadbearing capacity only) and which provides less than 15 min fire resistance (insulation).

wet rising main a rising main normally kept permanently charged with water.

wet sprinkler system a sprinkler system where the pipework is kept permanently charged with water.

Index

Page numbers printed in **bold** type refer to figures; those in *italic* to tables. References to entries in the glossary of terms are indicated by an asterisk (*) following the page number.